Y0-CFS-926

DISCARD

Applied Science Review™

# Ecology

Applied Science Review™

# Ecology

Carl R. Pratt, PhD
Associate Professor of Biology
College of New Rochelle
New Rochelle, N.Y.

Springhouse Corporation
Springhouse, Pennsylvania

# Staff

**EXECUTIVE DIRECTOR, EDITORIAL**
Stanley Loeb

**SENIOR PUBLISHER, TRADE AND TEXTBOOKS**
Minnie B. Rose, RN, BSN, MEd

**ART DIRECTOR**
John Hubbard

**ACQUISITIONS EDITOR**
Maryann Foley

**EDITORS**
Diane Labus, David Moreau, Kevin Law, Janice Fisher

**COPY EDITORS**
Diane M. Armento, Pamela Wingrod

**DESIGNERS**
Stephanie Peters (associate art director), Matie Patterson (senior designer), Donald G. Knauss

**ILLUSTRATORS**
Jacalyn Facciolo, Rhonda Forbes, Jean Gardner, John Gist, Judy Newhouse, Stellarvisions

**MANUFACTURING**
Deborah Meiris (director), Anna Brindisi, Kate Davis, T.A. Landis

**EDITORIAL ASSISTANTS**
Caroline Lemoine, Louise Quinn, Betsy K. Snyder

**Cover:** *Aerial view of fires in tropical rain forest.* Scott Thorn Barrows.

©1995 by Springhouse Corporation, 1111 Bethlehem Pike, P.O. Box 908, Springhouse, PA 19477-0908. All rights reserved. Reproduction in whole or part by any means whatsoever without written permission of the publisher is prohibited by law. Authorization to photocopy items for personal use, or the internal or personal use of specific clients, is granted by Springhouse Corporation for users registered with the Copyright Clearance Center (CCC) Transactional Reporting Service, provided the base fee of $00.00 per copy plus $.75 per page is paid directly to CCC, 27 Congress St., Salem, MA 01970. For those organizations that have been granted a license by CCC, a separate system of payment has been arranged. The fee code for users of the Transactional Reporting Service is 0874346894/95 $00.00 + $.75.
Printed in the United States of America.
ASR12-010794

> Library of Congress Cataloging-in-Publication Data
> Pratt, Carl Richard.
>   Ecology / Carl R. Pratt.
>     p. cm. – (Applied science review)
>   Includes bibliographical references and index.
>   1. Ecology – Outlines, syllabi, etc. I. Title II. Series.
> QH541.235.O98P73  1995
> 574.5 – dc20                                94-10645
>   ISBN 0-87434-689-4                           CIP

# Contents

*Advisory Board and Reviewers* . . . . . . . . . . . . . . . . . . . . . . . . . . vi
*Dedication* . . . . . . . . . . . . . . . . . . . . . . . . . . . . . . . . . . . . . . . vii
*Preface* . . . . . . . . . . . . . . . . . . . . . . . . . . . . . . . . . . . . . . . . viii

1. **Ecology as a Science** . . . . . . . . . . . . . . . . . . . . . . . . 1
2. **Ecosystem Basics** . . . . . . . . . . . . . . . . . . . . . . . . . . 6
3. **Soils** . . . . . . . . . . . . . . . . . . . . . . . . . . . . . . . . . . 23
4. **Energy Flow** . . . . . . . . . . . . . . . . . . . . . . . . . . . . . 31
5. **Movement of Materials Through Ecosystems** . . . . . . . . . 38
6. **Population Ecology Basics** . . . . . . . . . . . . . . . . . . . . 52
7. **Competition and Other Population Interactions** . . . . . . . . 63
8. **Predation** . . . . . . . . . . . . . . . . . . . . . . . . . . . . . . . 74
9. **Life History Patterns** . . . . . . . . . . . . . . . . . . . . . . . . 86
10. **Coevolution and Mutualism** . . . . . . . . . . . . . . . . . . . 95
11. **Community Ecology Basics** . . . . . . . . . . . . . . . . . . . 101
12. **Succession** . . . . . . . . . . . . . . . . . . . . . . . . . . . . . 112
13. **Disturbance in Ecology** . . . . . . . . . . . . . . . . . . . . . 126
14. **Terrestrial Ecosystems** . . . . . . . . . . . . . . . . . . . . . 134
15. **Freshwater Ecosystems** . . . . . . . . . . . . . . . . . . . . 140
16. **Marine Ecosystems** . . . . . . . . . . . . . . . . . . . . . . . 151

**Appendix: Glossary** . . . . . . . . . . . . . . . . . . . . . . . . . . . 159

**Selected References** . . . . . . . . . . . . . . . . . . . . . . . . . 165

**Index** . . . . . . . . . . . . . . . . . . . . . . . . . . . . . . . . . . . . 166

## Advisory Board

Leonard V. Crowley, MD
    Pathologist
    Riverside Medical Center
    Minneapolis;
    Visiting Professor
    College of St. Catherine, St. Mary's
      Campus
    Minneapolis;
    Adjunct Professor
    Lakewood Community College
    White Bear Lake, Minn.;
    Clinical Assistant Professor of Laboratory
      Medicine and Pathology
    University of Minnesota Medical School
    Minneapolis

David W. Garrison, PhD
    Associate Professor of Physical Therapy
    College of Allied Health
    University of Oklahoma Health Sciences
      Center
    Oklahoma City

Charlotte A. Johnston, PhD, RRA
    Chairman, Department of Health
      Information Management
    School of Allied Health Sciences
    Medical College of Georgia
    Augusta

Mary Jean Rutherford, MEd, MT(ASCP)SC
    Program Director
    Medical Technology and Medical
      Technicians—AS Programs;
    Assistant Professor in Medical Technology
    Arkansas State University
    College of Nursing and Health Professions
    State University

Jay W. Wilborn, CLS, MEd
    Director, MLT-AD Program
    Garland County Community College
    Hot Springs, Ark.

Kenneth Zwolski, RN, MS, MA, EdD
    Associate Professor
    College of New Rochelle
    School of Nursing
    New Rochelle, N.Y.

## Reviewers

Lorraine Mineo, BS, MA
    Lecturer
    Lafayette College
    Department of Biology
    Easton, Pa.

Neil Cumberlidge, PhD, PGCEd
    Associate Professor of Biology
    Northern Michigan University
    Marquette

Harry N. Cunningham, Jr., PhD
    Associate Professor of Biology
    Pennsylvania State University;
    Behrend College
    Erie, Pa.

**Dedication**

To my family for all their love and support.

# Preface

This book is one in a series designed to help students learn and study scientific concepts and essential information covered in core science subjects. Each book offers a comprehensive overview of a scientific subject as taught at the college or university level and features numerous illustrations and charts to enhance learning and studying. Each chapter includes a list of objectives, a detailed outline covering a course topic, and assorted study activities. A glossary appears at the end of each book; terms that appear in the glossary are highlighted throughout the book in boldface italic type.

*Ecology* provides conceptual and factual information on the various topics covered in most introductory ecology courses and textbooks and focuses on helping students to understand:
- the basic components and processes governing ecosystems
- the classifications and properties of soil
- energy flow throughout the ecosystems
- the nature of biogeochemical cycles
- properties, growth, and regulation of populations
- competition, mutualism, and commensalism among species
- life cycles and patterns
- the development of communities
- theories and models of succession
- the impact of ecological disturbances
- the characteristics of terrestrial and aquatic ecosystems.

# 1

# Ecology as a Science

## Objectives

After studying this chapter, the reader should be able to:
- Define ecosystem, environment, population, community, biome, and biosphere.
- Explain the difference between a theory, a hypothesis, descriptive studies, and experimentation.
- Describe the difference between a dependent variable and an independent variable.
- Explain the significance and importance of using models in ecology.

## I. Scope of Ecology

### A. General information
1. *Ecology* is the study of organisms (individuals or groups) in nature and their interrelationships and interactions between one another and their environment
2. In ecology, the *environment* includes all physical and biological aspects affecting the survival, growth, and reproduction of organisms; it includes such parameters as temperature, moisture, sunlight, and nutrients, as well as interactions between and among organisms
3. Ecology involves the study of the structure and function of nature
   a. Ecological structure involves the distribution and abundance of organisms
   b. Ecological function involves the study of the growth and interaction of organisms

### B. Hierarchy and levels of ecology
1. Ecologists study four natural units or levels: individuals, populations, communities, and ecosystems
   a. Ecologists who study individual members of a species examine their physiology and behavior in order to explain the species' survival, distribution, and abundance
   b. Ecologists who study a **population** examine collections of individuals of the same species in a particular geographical area and focus on how members of the population respond to one another and to their environment
   c. Ecologists who study a **community** examine all of the populations of organisms (plants, animals, and decomposers) that exist together in a given locale

- (1) An oak forest is an example of an ecological community that contains many populations of grasses, shrubs, herbs, trees, birds, mammals, insects, reptiles, bacteria, and fungi
- (2) At the community level, ecologists study the interaction between different populations; this interaction includes how the populations coexist and relate to one another in terms of obtaining items necessary for survival, such as food, shelter, and mates
- (3) Ecological communities are not static but change over time in response to climatic changes, human intervention, and geological events
- d. Ecologists who study an *ecosystem* examine an ecological unit that includes the community of living things interacting with one another and their physical environment
  - (1) An ecosystem is considered a self-sustaining, self-perpetuating, self-controlling collection of living and nonliving components
  - (2) Major topics of interest and study at the ecosystem level include the cycling and recycling of essential nutrients within the ecosystem and the flow of energy through the ecosystem
  - (3) Ecosystems generally are characterized by the community present; examples include ponds, streams, oceans, forests, and grasslands
- 2. In addition to studying the four natural units (individuals, populations, communities, and ecosystems), ecologists sometimes describe larger, less well defined units
  - a. Ecosystems that extend over large geographic areas (often vast portions of continents) are known as *biomes*
  - b. When taken as a large interacting unit, all of the ecosystems of the earth are known as the *biosphere* or ecosphere
    - (1) At the biosphere level, ecologists study global movements of energy and materials and the global patterns of distribution and migration of organisms
    - (2) The biosphere is considered to exist at the intersection of the *hydrosphere* (water bodies), the *atmosphere* (less than 7 miles above the earth's surface), and the *lithosphere* (soil and upper earth crust)
- 3. The levels of organization in ecology follow a hierarchical order or progression from the smallest unit, the individual, to the largest unit, the biosphere
  - a. Each level of organization includes a larger variety of organisms and, often, a greater expanse of the planet
  - b. With each higher level of organization, there also is an increase in the complexity and intensity of interactions among the various organisms and their environment

## II. Experimentation in Ecology

**A. General information**
1. Until the 1960s, ecological studies were largely descriptive; subsequently, description has been replaced with the experimental or empirical approach
   a. Early descriptive ecological research consisted of studies that characterized animal behavior, cataloged species in a particular area of interest, or described the pattern of community development in a particular locale
   b. More recently, ecologists have emphasized experimentation — trials to test or demonstrate particular ecological phenomena

c. Experimentation involves a more active manipulation of organisms or their environment by the researcher than does descriptive ecology, in which the researcher plays the role of a passive observer
2. The study of ecology is based on a collection of theories that form the foundation of our understanding of the natural world
  a. A *theory* is a statement or explanation of cause and effect that seems plausible but cannot be directly confirmed or rejected by experimentation
  b. Theories are interpretations of nature based on the conclusions of many experiments; they may be modified or rejected if new data or information that contradicts the prevailing theories becomes available
  c. Theories are important because they permit the formulation of testable hypotheses or questions regarding natural phenomena
  d. A key to the advancement of science in general, and ecology in particular, is people asking questions about nature and collecting evidence or data to answer those questions

## B. Hypothesis testing
1. A *hypothesis* is a statement about cause and effect that can be tested experimentally; it formulates part of a general theory that is to be tested
2. A hypothesis reflects past experience and knowledge regarding similar questions; it is based on what is already known
3. A hypothesis must be stated in terms that allow experimental testing
  a. A statement or idea stated in terms that permit testing by experimentation is considered a "testable" hypothesis
  b. A hypothesis can be disproved, falsified, or rejected, but it cannot be confirmed or proven with absolute certainty by experimentation
  c. To test a hypothesis, scientists perform experiments, collect data, and evaluate the results of those experiments
  d. Testing a hypothesis often involves some numerical or mathematical analysis
4. As the result of testing and evaluating results from experiments generated by one hypothesis, other hypotheses may be generated

## C. Laboratory and field studies
1. Testing a hypothesis involves collecting data by experimentation
2. Observational or descriptive studies are limited because mere observed similarities among phenomena do not provide evidence of cause and effect and do not allow for rigorous hypothesis testing
  a. Descriptive studies are those in which the researcher is merely an observer and does not actively manipulate or alter the situation being examined
  b. Observational or descriptive studies are valuable first steps in discovering and understanding natural occurrences; however, they generally do not provide sufficient understanding of a particular phenomenon
3. Experimentation involves active manipulation of systems and, usually, simplification because of the many variables involved
  a. In an experiment, the researcher alters or manipulates some factor or factors while holding others constant in order to detect any change
  b. Because an experiment can quickly become too complicated or unwieldy, an experimenter generally will simplify it by examining only one or a few factors at a time

4. Experiments may be performed in a laboratory under controlled conditions or in the field under more natural conditions
5. A controlled experiment involves perturbations — changes or alterations produced by the experimenter, who controls many factors while allowing one (or a few) variables of specific interest to change
   a. There are two types of variables in a controlled experiment: the experimental or independent variable that is being tested, and the dependent variable that is the result or change observed
      (1) The *dependent variable* depends on the value assigned to the independent variable
      (2) The *independent variable* is the factor that is being studied and is manipulated by the experimenter
      (3) For example, in an experiment examining plant growth (measured as weight) and fertilizer application, the amount of fertilizer applied is the experimental or independent variable and the weight of the plants is the dependent variable; the growth of the plants *depends* on the level of the independent variable — the amount of fertilizer
   b. A controlled experiment contains a control sample, which is subjected to all the steps and manipulations of the experiment except the one factor being tested (that is, the control group is not subjected to the experimental factor)
   c. In the previous example, an experimental group of plants would receive various amounts of the fertilizer, while the control group would receive the same care (light, water, temperature) but no fertilizer application
   d. The control sample is used as a *benchmark* or standard for comparison in order to detect differences among the experimental groups
6. A field experiment involves the manipulation of one or a few independent variables in natural communities; the results of field experiments often are hard to interpret because the researcher may have difficulty conducting a strictly controlled experiment
7. Because the conditions within a laboratory may be very different from conditions in nature, laboratory experiments often must be repeated in the field for verification and to ensure that laboratory results apply equally well to the natural world

## D. Inductive and deductive methods
1. Experimentation and scientific method involve two basic types of reasoning: inductive reasoning and deductive reasoning
2. *Inductive reasoning* takes specific knowledge and expands its application to a wider, more general set of concepts; it involves the investigation of correlation or association between classes of facts
   a. Inductive reasoning involves inferring generalities from a particular collection of facts and reasoning from a set of specific observations to reach a general conclusion
   b. The inductive approach entails the collection of facts, observations, or data; statistical analysis to discern relationships among the data; and formulation of conclusions regarding the hypothesis
   c. A major weakness of the inductive approach is that it can indicate only a correlation or relationship between occurrences; it does not establish cause and effect
3. *Deductive reasoning* (sometimes referred to as the hypothetico-deductive method) proceeds from the general to the specific and involves the development of a

general statement or research hypothesis first, followed by collection of data to support or refute the statement
   a. In deductive reasoning, the reasoning flows in the opposite direction from that in inductive reasoning; in deduction, specific results are extrapolated from general premises or principles
   b. Deductive reasoning commonly is used in ecology
   c. Deductive reasoning begins with an "if . . . then" statement, such as "If plant nutrients are important to the growth and health of plants, then it should be possible to find evidence that varying nutrient availabilities will result in variations in plant growth and health"
   d. In deductive reasoning and subsequent hypothesis testing, it is critical that controlled experiments be designed carefully to test the specific hypothesis of interest

### E. Models in ecology
  1. A *model* is a representation and simplification of a natural phenomenon developed to predict a new phenomenon or to provide insights into existing phenomena; models may be verbal, written, graphic, or mathematical explanations
  2. Once developed, models are validated and verified with observational studies and experimental data
      a. *Validation* is the explicit and objective testing of a basic hypothesis, accomplished by measuring the extent to which the results of the model agree with the behavior of the real-life system
      b. *Verification* is the process of testing whether or not the model is a reasonable representation of the real-life system being investigated and if its parts or components agree or coincide with known mechanisms of the system
  3. Models are important devices to help ecologists organize their understanding of natural phenomena; they are important because often they serve as mechanisms to generate new hypotheses for testing

# Study Activities
1. Write a brief paragraph describing the relationships among populations, communities, and ecosystems.
2. Formulate a simple experiment; indicate your hypothesis, dependent variables, independent variables, and control and experimental groups.
3. List, in hierarchical order, the natural units studied by ecologists.
4. Give an example of a population of organisms in a pond or lake.
5. A city park can be considered to support a community of organisms. What populations are present?
6. Explain why ecologists use models in their research.

# 2

# Ecosystem Basics

## Objectives

After studying this chapter, the reader should be able to:
- Define and describe an ecosystem in terms of its components, inputs, and outputs.
- Describe the difference between acclimatization and adaptation to environmental change.
- Explain the concepts of tolerance range and limiting factors.
- Describe the importance of solar radiation in determining global patterns of wind, ocean currents, and climate.
- Explain some of the responses of plants and animals to varying moisture conditions in their environment.
- Describe plant and animal adaptations to high and low temperatures.
- Explain the importance of light in determining the periodicity of activities in plants and animals.
- Explain the role of nutrients in determining the distribution of plants and animals.

## I. The Concept of the Ecosystem

### A. General information
1. Living organisms and their nonliving environment are inseparably interconnected and intertwined
   a. Together, the living organisms are termed **biotic** factors or components
   b. The nonliving entities of the environment, such as temperature, moisture, air, soil, nutrients, and light, are termed *abiotic* factors or components
2. An *ecosystem* is a collection of all the living (biotic) organisms and their physical environment interacting with each other
3. Ecosystems are self-sustaining and self-regulating collections of biotic and abiotic components (for example, a pond, an ocean, a forest) and often have internal mechanisms that control the relationships among the various components

### B. Components of ecosystems
1. Ecosystems have boundaries or edges — for example, the edge of a forest or the edge of a pond (see *Ecosystem Structure*)
2. Ecosystems can be characterized by a variety of *inputs* (items or organisms entering an ecosystem), *outputs* (items or organisms exiting an ecosystem), and *components* (items or organisms that comprise the ecosystem)
   a. Components, inputs, and outputs may be either biotic or abiotic

## Ecosystem Structure

An ecosystem is a self-sustaining collection of living and nonliving components. Major inputs include solar energy, nutrients, water, carbon dioxide, and oxygen. Major outputs include losses consisting of water, nutrients, carbon dioxide, and oxygen, as well as losses due to heat and respiration.

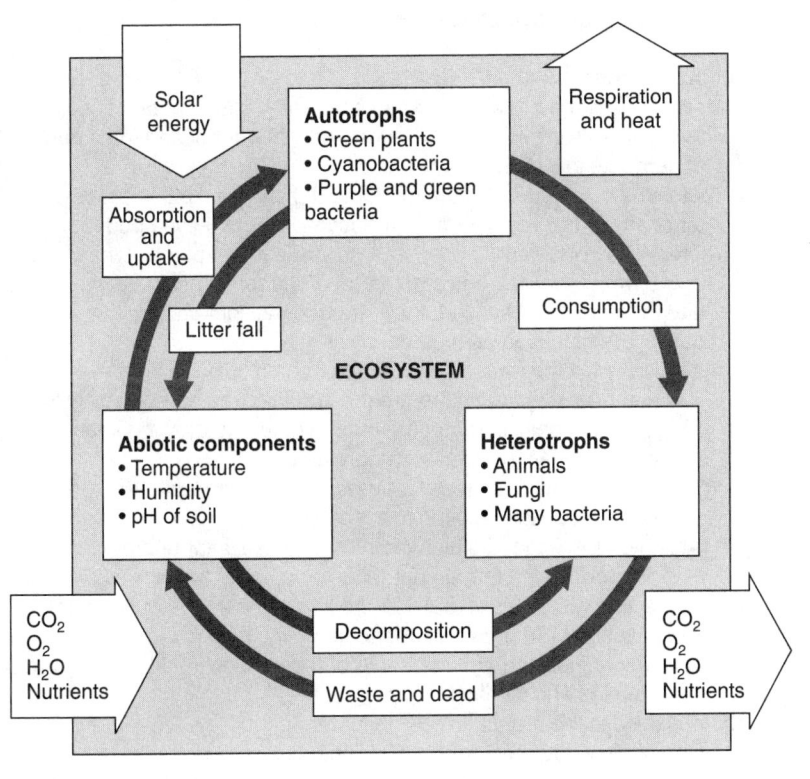

b. Biotic components are classified according to how they obtain their food
   (1) **Producers,** or **autotrophs,** are organisms that can synthesize their own foodstuffs; major producers are algae and plants that absorb sunlight and use its energy to synthesize organic foodstuffs from water and carbon dioxide in a process known as photosynthesis
   (2) **Consumers,** or **heterotrophs,** are organisms that require a supply of organic matter or food from the environment; they typically must "eat" other organisms to survive
      (a) Organisms that consume plant tissue are called *herbivores*
      (b) Organisms that consume animal tissue are called *carnivores*
      (c) Organisms that consume a combination of plant and animal tissues are called *omnivores*
      (d) Organisms that consume largely dead or decaying material and convert it into inorganic substances are called *decomposers* or *saprophytes*

(e) Organisms that live in association with and at the expense of other organisms from which they obtain organic nutrition are called *parasites*

   c. Abiotic components include energy (sunlight); physical factors (such as temperature, light, and humidity); and chemical factors, both inorganic ions (for example, iron, magnesium, sodium, chlorine) and organic molecules (proteins, carbohydrates, lipids, and nucleic acids)

## C. Essential processes

1. *Photosynthesis* is the metabolic process by which plants obtain energy
   a. Photosynthesis involves the capturing of light energy and the incorporation of nutrients into active plant tissues
   b. Photosynthesis serves as the fundamental energy fixation process of ecosystems and is the basis for the production of organic compounds
   c. Photosynthesis is carried out in the green tissue of plants, typically the leaves
   d. The *leaf area index* (LAI) is a measure of the relationship between leaves and the amount of sunlight they intercept as light passes through the various layers of leaves on a plant
      (1) The LAI is the ratio of leaf area per unit of ground area — the amount of leaf area divided by the ground surface area covered by the plant
      (2) The LAI usually is low at the beginning of the growing season and increases as the season progresses
      (3) A low LAI indicates less leaf area than ground area and represents wasted sunlight striking the ground
      (4) A high LAI indicates high maintenance costs for unproductive leaves; shaded leaves that do not receive sufficient light for efficient photosynthesis will be a net drain on the energy reserves of the plant
      (5) An optimal LAI depends on the intensity of light, the shape and arrangement of leaves, and the angle of the sun; optimal LAI varies with each species and climatic region
2. *Decomposition* is the reduction by consumers of energy-rich organic matter to carbon dioxide, water, and inorganic nutrients
   a. Decomposition involves the processes of fragmentation, mixing, and changes in the physical and chemical structure of the dead material
   b. Bacteria, fungi, and many small invertebrate organisms are important decomposer organisms
   c. In terrestrial systems, decomposition generally occurs in the upper soil layers
   d. Decomposition is "natural recycling," the conversion of organic materials to inorganic substances, which then are released into the ecosystem to be taken up by other organisms as nutrients

# II. Tolerance Range, Accommodation, and Limiting Factors

## A. General information

1. Organisms are subjected to a range of environmental factors during their lifetimes; for example, temperature may vary widely, light conditions change, and moisture conditions can be variable

2. Every organism has limits as to the extremes of abiotic factors that it can survive; for example, extreme high or low temperatures or extremely dry conditions may preclude the survival of an organism in a particular location
3. To examine the occurrence or distribution of an organism, ecologists examine an organism's tolerance range or set of physical factors within which an organism can survive and the limiting factors that constrain or limit the organism's numbers or distribution

## B. Tolerance range
1. The *tolerance range* of an organism is the spectrum of life requisites or requirements for sustaining life (physical factors) over which an organism can survive, reproduce, and persist over time
2. Organisms generally will not survive equally well over an entire range of tolerance; typically, there is an optimum range for survival and upper or lower limits for survival
3. An organism that displays a broad tolerance range for a particular factor often is described by the prefix "eury-"; for example, an organism with a broad tolerance range for heat is *eurythermal*, and one with a broad tolerance range for salinity is *euryhaline*
4. An organism that displays a narrow tolerance range for a particular factor often is described by the prefix "steno-"; for example, an organism with a narrow tolerance range for heat is *stenothermal*, and one with a narrow tolerance range for salinity is termed *stenohaline*

## C. Accommodation
1. The environment an organism encounters often changes with time; the ability to respond to or accommodate these changes may occur at the level of an individual organism or at the level of a population
2. If the accommodation to change in an environmental condition is by the entire species, it occurs over a number of generations in response to natural selection and results in a change in *gene frequency;* then it is termed **adaptation**
3. If the accommodation to change occurs at the individual level, it is characterized by changes in tolerance range over a short-term (seasonal) basis and occurs in response to changes in the natural environment; then it is termed **acclimatization**
   a. Acclimatization often is temporary and cyclic
   b. Many fish acclimatize to changes in their thermal environment during changes in season
   c. The lethality of different temperature extremes changes as the temperature range to which an organism is acclimatized changes; fish acclimatized to cold water in winter would be killed by an abrupt rise in temperature while a similar rise in temperature might not be lethal in the summer when they are acclimatized to warmer temperatures

## D. Limiting factors
1. To exist and thrive in a given environment, an organism must have all essential elements and materials for growth and reproduction
2. Basic requirements for survival, growth, and reproduction vary from organism to organism

3. Any factor or condition that approaches or exceeds an organism's tolerance range is considered a *limiting factor*
4. *Liebig's law of the minimum* (named after Justus von Liebig, 1840) states that plant growth is directly proportional to the supply of the nutrient present in minimum amount; the nutrient in minimum amount will tend to limit or restrict growth of the plant
   a. Liebig's law is strictly applicable only under steady-state conditions — that is, when energy and material flow into and out of the system in balance
   b. Organisms sometimes can compensate for limiting factors by *factor interaction,* a situation in which a high concentration or availability of one factor may modify the rate of utilization of another factor (for example, plants growing in shade require less zinc as a mineral nutrient than plants growing in full sunlight)
5. *Shelford's law of tolerance* (named after V.E. Shelford, 1913) states that the absence or failure of an organism can be controlled by a qualitative or quantitative deficiency or excess with respect to any one of several factors that may approach the organism's limits of tolerance
   a. According to Shelford, the distribution or occurrence of an organism can be controlled by an overabundance or deficiency of a particular factor
   b. An organism may have a wide tolerance range for one factor and a narrow tolerance range for another factor
   c. Organisms with wide tolerance ranges for all factors are likely to be widely distributed
   d. When conditions are not optimum for an organism with respect to one factor, the limits of tolerance for other factors may be reduced — a condition termed *factor compensation*
   e. Frequently, in nature an organism may not be living at the optimum of its tolerance range for a particular factor; in such cases, some other factor not being considered may be of greater importance
   f. The period of reproduction usually is a critical time when organisms are very sensitive to changes in environmental factors, and it is a time when tolerance ranges are very critical
   g. All physical requirements may be within the tolerance range for an organism, yet the organism still may fail as a result of biological interactions, such as competition
   h. Narrow tolerance ranges may be considered a form of specialization
   i. Conditions need not be perpetually outside an organism's tolerance range for that organism to be absent; for example, a factory spill into a stream might occur infrequently, but if the discharge exceeds an organism's tolerance range, then that organism may be absent from the area

## III. Climate

### A. General information
1. *Climate* is defined as the average weather conditions in an area over a period of years; it includes trends in temperature, precipitation, and wind velocity
2. Climate determines the availability of water and the overall temperatures, thus influencing the development of soils, the type of vegetation, and the composition of the biological community in an area

3. Because climate influences temperature and moisture, it also influences nutrient cycling, photosynthesis, and decomposition — all of which are temperature- and moisture-dependent

## B. Global patterns
1. The major determinant of climate is input from the sun, or *solar radiation*
   a. Differences in the amount of solar radiation received are responsible for variations in the heating of different parts of the planet; this uneven heating, coupled with the rotation of the earth, generates wind and ocean currents
   b. Approximately 32% of the solar radiation reaching the earth from the sun is reflected back into space by the earth's atmosphere
   c. Of the solar radiation not reflected, about 3% is absorbed by clouds and 15% is absorbed by dust, water vapor, and carbon dioxide in the atmosphere
   d. As a result of absorptive and reflective losses, only about 50% of the solar radiation reaching the outer layers of the atmosphere passes through the atmosphere to the earth's surface
   e. As solar radiation passes through the atmosphere, certain wavelengths are absorbed, so that the character of the solar radiation is changed; for example, nearly all of the ultraviolet-B wavelengths are absorbed by the ozone layer in the atmosphere so that virtually none reaches the earth's surface
   f. Solar radiation absorbed as short-wave heat energy (infrared) by the atmosphere and re-radiated as longer wave radiation is largely responsible for heating the earth's surface
   g. The percentage of solar radiation reflected back out into space by the earth's surface is called the *albedo*
      (1) The albedo determines how quickly and to what degree the earth's surface is heated
      (2) Water bodies have a variable albedo: low albedo when light strikes the water surface perpendicularly (approximately 2% is reflected), and high albedo when the sun is low on the horizon and light strikes the water surface at an angle (approximately 75% to 90% is reflected)
      (3) The albedo for snow and ice is 40% to 90%; dark forests and grasslands have albedos of 5% to 30%
      (4) Albedo varies from one region to another and is partly responsible for differences in surface heating of the planet
2. Sunlight strikes the equatorial regions of the earth more directly than the polar regions; as a result, the lower latitudes are heated more than the polar ones
3. The unequal heating of the earth by the sun results in global patterns of air circulation within the atmosphere
   a. Air heated at the equator rises until it reaches the stratosphere, where it cools and then spreads north and south toward the poles
   b. As the air masses reach the poles, they cool further, becoming denser and heavier and sinking over the arctic and antarctic regions
   c. The cool, heavier air reaches the earth's surface and flows toward the equator to replace the warmer air that is rising
   d. The general flow of air from the equator to the poles and back is altered by the earth's west-to-east rotation
   e. Air masses near the equator veer in the direction of the earth's rotation

f. The combination of the uneven heating of the atmosphere coupled with the earth's rotation results in the series of belts of prevailing east winds known as the *trade winds* near the equator and the *polar easterlies* at the poles
   g. In the middle latitudes is a region of west winds known as the *westerlies*
   4. Land masses heat and cool more rapidly than oceans, resulting in thermal differences that can cause wind circulation patterns
   5. The wind circulation patterns, and the temperature and moisture regimes they generate, are in large part responsible for the climatic regions of the planet and thereby influence the distribution of forests, grasslands, and deserts
   6. In a pattern analogous to global air flow, the earth's rotation, solar energy, and winds produce the major ocean currents
      a. Land masses divide the oceans and deflect some ocean currents
      b. The intense sunlight at the equator produces warm, expanding water in the tropical oceans that tends to flow northward
      c. Wind patterns, together with the earth's rotation, distort the northward flow of water, resulting in the circulation of currents in a series of cells in the two hemispheres
      d. Mainly, circulation of water just north of the equator is in a clockwise or westward direction; it changes to an eastward-moving current at approximately 40° N latitude
      e. Just south of the equator, the water flow is westward; it changes to an eastward flow from 40° S latitude to the Antarctic
      f. Higher temperatures in the tropics result in higher evaporation from the oceans and, consequently, higher salinity in tropical waters as compared to temperate waters
      g. Ocean currents can influence coastal region ecosystems; for example, the Gulf Stream brings milder temperatures and more moisture to Great Britain and Norway than would otherwise occur at their latitudes

## C. Regional climates
   1. Mountain ranges, bodies of water, and other geographical features can affect local or regional climates
   2. Mountain ranges influence regional climates by changing patterns of precipitation
      a. As an air mass reaches a mountain, it ascends, cools, and (because cool air holds less water than warm air) releases precipitation on the windward side of the mountain
      b. As the cool, dry air descends the leeward side of the mountain, it warms and picks up moisture but releases little precipitation
      c. The result is lush vegetation on the windward side of the mountain and dry, sometimes desertlike conditions on the leeward side
   3. The regional climate of areas adjacent to large bodies of water is tempered by the thermal inertia of lakes; such regions will have somewhat milder winter temperatures, because the lake water is slow to cool and ameliorates the adjacent air temperatures

## D. Microclimates
   1. Variation in temperature, moisture, air movements, and light intensity within a particular area is known as *microclimate*

2. The vegetation cover influences the microclimate of an area, especially near the ground, by altering wind movements, evaporation rates, moisture levels, and soil temperatures
3. A cluster of vegetation deflects winds up and over its top, creating very different microclimate conditions within a patch of vegetation than would be encountered in the area surrounding the patch
4. Large microclimatic differences exist between north-facing and south-facing slopes as a result of the amount of direct sunlight received and the length of periods of shading
    a. In the Northern Hemisphere, south-facing slopes receive more solar energy
    b. South-facing slopes are characterized as having higher temperatures, up to 50% higher rates of evaporation, lower soil moisture, and more variable extremes in these parameters
    c. Differences in microclimate result in different plant communities on south-facing and north-facing slopes
    d. Because animals are mobile, there are few if any differences between those found on south-facing and north-facing slopes

# IV. Moisture

## A. General information
1. Water is a crucial component of living things and is required in some form by all living things
2. Water is found in various locations and physical states on the earth

## B. Water or hydrologic cycle
1. Water is distributed globally in oceans, lakes, streams, atmosphere, and soil
    a. Approximately 97.6% of the planet's water resources are considered saltwater (oceans comprise 71% of the earth's surface)
    b. The remaining 2.4% of the world's water resources are very low in salts and considered freshwater (freshwater is located in streams, lakes, ponds, rivers, ice packs, glaciers, soil moisture, and the atmosphere)
2. Water moves from one compartment (for example, ocean, lake, atmosphere, or soil) to another in a process referred to as the *water* or *hydrologic cycle,* which consists of three major stages: evaporation, condensation, and precipitation (for more information on this cycle, see section I. in Chapter 5, Movement of Materials Through Ecosystems)

## C. Humidity
1. Water vapor in the atmosphere generally is measured as the *relative humidity* (RH), defined as the percentage of water in the air relative to the amount of water that the air could hold were it fully saturated at a given temperature
    a. Warm air holds more moisture (has a higher RH) than cool air
    b. RH varies according to differences in topography, vegetation, elevation, and microclimate
2. Temperature and wind both affect evaporation from plant, soil, and other environmental surfaces and thus affect relative humidity

## D. Plant responses to moisture
1. Plants can be considered poikilohydric or homoiohydric in their responses to changes in water availability
2. *Poikilohydric plants* lack mechanisms to regulate their overall water content; thus, their water status tends to reflect available environmental moisture conditions
   a. Poikilohydric plants typically confine their growth to moisture periods; during dry periods, they often become dormant
   b. Examples of poikilohydric plants include algae, fungi, lichens, and mosses
3. *Homoiohydric plants* possess mechanisms to control the water content of their bodies; thus, their water status is more or less independent of environmental conditions
   a. Some of the adaptations possessed by homoiohydric plants include large water vacuoles in cells, a thick impermeable cuticle covering of the epidermal surfaces, fewer stomata, stomata sometimes located in pits, and an extensive root system for extracting water from soil
      (1) *Water vacuoles* are membrane-enclosed sacs within mature plant cells that contain water
      (2) *Stomata* are microscopic pores in the outer surface of plant leaves and stems that allow gas exchange between the environment and the plant's interior
   b. Examples of homoiohydric plants include ferns, evergreen trees, and flowering plants
4. Many plants regularly must face periodic times of little or no water (seasonal drought)
   a. Plants are considered *drought-resistant* if they are able to withstand lack of water
   b. Drought resistance involves two components: drought tolerance and drought avoidance
      (1) *Drought tolerance* is the ability of a plant to maintain its physiological activity despite a lack of water or the ability to survive drying of tissues
      (2) *Drought avoidance* involves adaptations that allow the plant to avoid, oppose, or ameliorate the dry conditions; some plants become dormant and produce seeds during the dry season, whereas other close their stomata to prevent water loss by transpiration
   c. Extended dry periods often stress plants and may weaken them, making them vulnerable to insect attack, fire, or disease that they would otherwise resist
5. Too much moisture, as in the case of flooding, results in a decline in exchange of the gases carbon dioxide and oxygen between the roots and the soil; for proper root growth and survival, oxygen must be obtained by the roots from the surrounding soil, and carbon dioxide (a waste product) must leave the roots and enter the surrounding soil
   a. Prolonged flooding often results in the death of plants caused by asphyxiation of their roots
   b. Wetland plants have evolved adaptations to deal with waterlogged soils
      (1) Many herbaceous species have gas-filled chambers in their stems and roots through which oxygen diffuses to the root cells
      (2) Some woody species, such as mangroves, produce *pneumatophores,* specialized roots that provide pathways for gas diffusion in the roots

### E. Animal responses to moisture
1. *Osmolarity* is defined as the total osmotic pressure developed by a mixed solution of chemicals (such as blood and tissue fluids of animals); differences in osmolarity between internal body fluids and the external environment will result in water movement (osmosis)
   a. Animals living in freshwater environments (with a low saline concentration) have body fluids with a higher total osmolarity than their surroundings (that is, they are *hyperosmotic*) and thus tend to gain water and lose salts
   b. Freshwater residents survive by opposing these exchanges of water and salts
2. Freshwater fish and freshwater invertebrates, such as crayfish, eliminate excess water by producing large amounts of dilute urine and replace their lost salts by actively pumping ions into the blood via the gills
3. Marine invertebrates are *iso-osmotic* and possess body fluids with a total osmolarity that is approximately equal to that of seawater; they have few osmotic or ionic problems
4. *Hypo-osmotic* marine fish have body fluids with a total osmolarity that is less than seawater; they tend to lose water and gain salts and oppose these tendencies by drinking seawater, producing small amounts of concentrated urine, and actively expelling excess salts across their gills
5. Marine birds (gulls and albatrosses) and reptiles (sea snakes and turtles) have salt glands or specialized tear glands that secrete excess salt from their bodies
6. Marine mammals (whales and dolphins) have very efficient kidneys; they do not drink water, but rather obtain it from food sources
7. Terrestrial animals gain water in three major ways: drinking, obtaining water from the fluids in their food, and producing water as a by-product of their metabolic pathways (cellular chemistry)
   a. Terrestrial animals generally are highly mobile and may avoid dry conditions (or flooding conditions) by migrating or by altering their activity patterns so as to avoid adverse conditions
   b. Most terrestrial mammals also possess very efficient kidneys that conserve water while eliminating excess salt and waste materials from the blood

## V. Temperature

### A. General information
1. Environmental temperature depends, in large part, on the amount of solar radiation reaching the habitat of plants and animals
2. The amount of solar radiation reaching any particular point on earth is influenced by such factors as season, aspect (orientation to the sun), time of day, and slope

### B. Thermal exchange
1. Living organisms must exchange energy with their environment and balance the amount of heat energy gained from the environment with heat energy lost to the environment in a process referred to as *thermal exchange*
2. For an organism to be in temperature balance, the amount of heat energy absorbed from the environment plus metabolic energy must equal the thermal energy lost to the environment

3. Heat gains and losses can occur through various modes such as conduction, convection, and evaporation
    a. A major source of heat energy absorption is solar radiation
    b. Environmental surfaces (for example, rocks, soils) as well as organisms absorb solar radiation and emit infrared radiation to their surroundings
    c. Heat also is transferred directly from one substance to another by the process of *conduction;* the amount of heat lost or gained by conduction depends on the surface area exposed and on the distance and temperature differences between the two surfaces
    d. *Convection* is the transfer of heat by the flow of a fluid (for example, water or air)
        (1) Convection may occur naturally around an organism, or it may be forced by the movement of the organism or some part of the organism
        (2) The amount of heat lost or gained by convection depends on the shape and area of the organism, the velocity of the fluid, and the physical properties of both the organism and the fluid
    e. *Evaporation* is the physical change of a liquid to a gas
        (1) Evaporation enables organisms to lower their body temperature by transferring heat to the water (liquid), causing it to change physical state and become a gas (water vapor)
        (2) Evaporation requires heat as an energy source; heat is drawn from the organism's body, producing a cooling effect

## C. Plant responses to temperature
1. *Heat stress* occurs when plants are subjected to high temperatures
    a. Photosynthesis is very sensitive to heat stress, because many of the controlling enzymes are heat-sensitive
    b. Most plants die if leaf, stem, or root temperatures reach 44° to 50° C
        (1) A plant's ability to tolerate high temperatures varies seasonally and with its developmental (life cycle) stage
        (2) For example, dry, dormant wheat seeds can withstand ambient temperatures of 90° C for 10 minutes and remain viable, but after imbibing water for 24 hours, exposure to 60° C for 1 minute is lethal
2. *Cold stress* occurs when plants are subjected to low temperatures
    a. When the ambient temperature drops below the minimum for growth, the plant becomes dormant even though photosynthesis and respiration may continue, though at slower rates
    b. At low temperatures, photosynthesis is restricted in all plants and ceases completely in more sensitive plants at 5° to 10° C
    c. If freezing occurs slowly, ice crystals may form outside of cells, drawing water out of cells and causing dehydration
    d. The underground parts of plants (roots, bulbs, and rhizomes) are least sensitive to freezing, with lethal temperatures ranging between −10° and −30° C
    e. Sugars and other solutes found in the sap and other fluids of cold-adapted plants serve to depress the freezing point of the fluids and help the plants resist frost damage
    f. Many plants are injured by chilling to 10° C caused by disruption of plasma membranes; chilling injury causes a malfunction in the plant's water uptake and retention

       g. Tolerance to cold temperatures can be genetically determined; thus, tropical and subtropical plants have little resistance to low temperatures
       h. Plants native to seasonally cold climates build up a tolerance to frost coupled with winter dormancy in a process referred to as *cold hardening*
          (1) Cold hardening occurs gradually in autumn as a response to moderate chilling (temperatures of 5° to 0° C)
          (2) The gradual temperature changes appear to stimulate the production of protective antifreeze compounds in cells (such as certain sugars and amino acids)
   3. *Thermogenesis* is the metabolic production of heat to counteract the loss of heat to a colder environment
       a. A few plants are capable of generating extra heat by bringing about increases in their metabolic rate (that is, they use their own cellular chemistry to generate heat)
       b. Thermogenesis generally is restricted to members of the arum lily family (jack-in-the-pulpits, philodendrons, calla lilies, and skunk cabbages)
       c. Thermogenesis is thought to be an adaptive advantage in some plants, allowing them to initiate growth early in spring and thus get a "head start"

## D. Animal responses to temperature
   1. Animals' responses to temperature changes are influenced by their physiology, morphology, and behavior, and the environmental conditions to which they are exposed
   2. Aquatic animals live in a more thermally stable environment than do terrestrial animals and generally have a lower tolerance for temperature change than do terrestrial species
   3. Because air can heat up or cool down more rapidly than water, air temperatures fluctuate more radically than water temperatures
   4. Physiologically, animals can be placed into three categories with respect to their thermal behavior: ectotherms, endotherms, and heterotherms
   5. **Ectotherms** are animals with constantly varying body temperatures; they derive their body heat almost entirely from external heat sources, primarily solar radiation
       a. Ectotherms include most invertebrates, reptiles, fish, and amphibians
       b. If environmental temperature fluctuates during a 24-hour period, the body temperature of an ectotherm also is likely to fluctuate; because of this, ectotherms also are called ***poikilotherms***
       c. Ectotherms often are referred to as "cold-blooded"; this term is misleading because their blood may be quite warm, approaching the temperature of so-called "warm-blooded" animals
       d. Ectotherms have a high thermal conductivity between the body and the environment and a low metabolic rate
       e. The metabolic rate of ectotherms varies according to temperature; it is low at low temperatures and rises as the temperature rises, at a rate that doubles for every 10° C rise in temperature, because rising temperatures increase the rate of enzymatic reactions that control metabolic rate
       f. Lacking the ability to regulate body temperature internally, ectotherms must use behavioral mechanisms to control body temperature
          (1) To increase body temperature, many ectotherms use *basking behavior,* in which they lie exposed to the sun to absorb solar radiation

(2) To lower body temperature, ectotherms may seek shade or the refuge of a burrow to escape periods of high ambient temperature
- g. Generally, aquatic ectotherms encounter smaller fluctuations in temperature and often are more sensitive to abrupt or sudden changes in temperature
- h. Ectotherms generally become lethargic and inactive at cool temperatures, but their food requirements also lessen
- i. Energy production in ectotherms largely is anaerobic (50% to 98% of it), and consequently they often deplete cellular energy rapidly and accumulate lactic acid in muscle tissues; this property severely limits bursts of poikilothermic activity to a few minutes before exhaustion sets in
- j. As a result of anaerobic muscle activity, ectotherms often are limited in activity level; thus, predaceous terrestrial ectotherms often secure prey by ambush rather than pursuit, for which they lack the stamina

6. **Endotherms** establish and maintain high and constant internal body temperatures by producing metabolic heat from their internal (metabolic) processes
   - a. Endotherms include birds and mammals
   - b. The body temperature of endotherms remains relatively constant; because of this, endotherms also are known as **homoiotherms** or **homeotherms**
   - c. Endotherms commonly are referred to as "warm-blooded"
   - d. In cold and temperate climates, the body temperature of endotherms is derived from metabolic processes, and heat loss to the environment is slowed by layers of insulation (fur, feathers, or subcutaneous fat or blubber) of varying thickness; excess heat is lost by evaporative cooling (sweating, fur licking, gular fluttering, or panting)
   - e. The maintenance of high body temperature (about 37° C for mammals, 40° C for birds) offers the advantage of high-performance activities regardless of environmental temperature; the disadvantage is that a high metabolic rate requires that homoiotherms eat large quantities of high-quality food every day
   - f. Endotherms use 80% to 90% of their energy budgets (input) to maintain body temperature

7. **Heterotherms** at times have high and well-regulated body temperatures but at other times undergo rapid, large-scale swings in their body temperatures, either over the short term on a daily basis (adult flying insects, bats, hummingbirds, sunbirds) or, over longer periods, on a seasonal basis (hibernators)
   - a. At some time during their daily or seasonal cycle or environmental situation, heterotherms take on the characteristics of both ectotherms and endotherms
   - b. Heterotherms include most adult flying insects (bees, butterflies, dragonflies, moths), hummingbirds, sunbirds, bats, ground squirrels, some species of mice (pocket mice, kangaroo mice, white-footed mice), woodchucks, and marmots
   - c. Heterotherms maintain homeostatic temperatures during much of their life, but when environmental parameters (temperature or moisture) become stressful, they allow a measured drop in body temperature, heart rate, and metabolic rate but still regulate their body temperatures around the new, lower set point to save energy or avoid adverse environmental conditions
   - d. "Warm-blooded" heterotherms are able to enter a period of reduced metabolic actively during which body temperature is greatly reduced, often near ambient; this is referred to as **torpor**

(1) Torpor is characterized by a loss of homoiothermy, reduction in metabolism, lower respiration and heart rate, decreased energy demands, and decreased kidney and digestive functions
(2) A deep state of torpor that occurs generally during the winter season, lasts several months, and appears to be a mechanism to avoid adverse conditions is called **hibernation** and occurs in animals such as mice, ground squirrels, woodchucks, and gophers
(3) **Estivation** is a state of torpor that occurs generally in the summer season, typically lasts hours or days, and appears to be a mechanism to avoid temperature and water stress; estivation is used by many desert animals
(4) Hummingbirds and sunbirds enter torpor overnight on a nearly daily basis to conserve energy

## E. Temperature and survival
1. Death at high temperature occurs as a result of enzyme inactivation, metabolic imbalance (for example, at high temperature, respiration in plants proceeds faster than photosynthesis), or dehydration
2. Carnivore food sources tend to contain more water than plant tissues; because mammalian carnivores often are very mobile and can travel to water, they have sufficient water to "spend" on thermoregulation (for example, they can use water for evaporative cooling)
3. Large desert herbivores (for example, eland, oryx [African antelope], and gazelle) can travel to shade and water; they can afford evaporative losses and often permit body temperature to rise to 42° C
4. The fur of daytime-active desert animals reflects solar energy and is thicker on dorsal surfaces; it is estimated that the thick, coarse hair on the back of a camel reduces evaporative water loss by 50%
5. Panting can enhance heat exchange by promoting evaporative cooling from the mouth
6. Many animals have arteries and veins divided into small parallel, intermingling vessels that form a vascular bundle known as the *rete*
   a. Within the rete, blood flows in opposite directions, and a heat exchange takes place
   b. This exchange between parallel vessels (arteries and veins) with opposing flow is known as *counter-current exchange*
   c. Counter-current heat exchange occurs within the rete of oryx and gazelles
      (1) The external carotid arteries pass through a sinus containing venous blood vessels filled with blood cooled by evaporation from the moist surfaces of the nasal cavities
      (2) Through the counter-current exchange mechanism, the blood in the carotid arteries is cooled to prevent the brain from overheating during periods of high ambient temperature or physical exertion
      (3) Arterial blood passing through the rete on its way to the brain is cooled by 2° to 3° C
   d. In many arctic mammals (for example, arctic fox and polar bear), counter-current exchange prevents excessive heat loss through appendages; the rete is located at the junction of the appendages to the body

## F. Temperature and species distribution
1. Temperature is an important determinant of species distribution
2. Generally, many species are restricted in their distribution by the lowest critical temperature of their life cycle, which usually is the reproductive period
3. In terrestrial ecosystems, it often is difficult to separate the influence of temperature and moisture on the distribution of a species because the two factors usually are intertwined

# VI. Light

## A. General information
1. Light affects the rate of photosynthesis, the local distribution of plants and animals, and the seasonal activities of organisms
2. Light is only one part of the spectrum of energy contained in solar radiation; generally, it is considered to be that portion of the spectrum from 400 to 700 nanometers in wavelength

## B. Plant adaptations to light
1. In terrestrial environments, vegetation modifies the light as it passes through the many layers of leaves and stems
   a. In both grasslands and forests, most of the light that enters these systems is intercepted by the leaf tissue of the plant species present
   b. As light passes downward through the many layers of leaves (known as the leaf canopy), its intensity declines and the composition of the light changes
      (1) In an oak forest, only 6% of the total midday sunlight reaches the forest floor; the rest is intercepted by the canopy
      (2) In a tropical rain forest, approximately 0.25% to 2% of sunlight reaches the forest floor
      (3) In a pine forest, about 10% to 15% of sunlight reaches the forest floor
      (4) In a typical grassland, only about 2% to 5% of sunlight reaches the soil surface
   c. Tree species vary in their ability to survive in the shade of other trees; the ability to survive in the shade is referred to as *shade tolerance*
      (1) Shade-intolerant species are adapted to high light intensities and have high rates of respiration
      (2) Shade-intolerant species can be considered shade avoiders and are found in open sites and disturbed areas; examples include jack pine, tamarack, willow, black locust, and birch
      (3) Shade-tolerant species do not compete with shade-intolerant species when in full sunlight and generally have lower rates of photosynthesis and lower respiration rates; as a result, they photosynthesize at lower light intensities
      (4) Shade-tolerant species exploit flecks of light that penetrate the canopy as the wind blows the branches above and may live suppressed in the shade of other plants, ready to grow rapidly if other trees are removed and more light reaches them; examples include beech, dogwood, sugar maple, white cedar, and hemlock
2. In aquatic environments, water modifies the light as it passes through

a. Approximately 10% of sunlight striking the surface of water bodies directly is reflected; sunlight hitting water bodies at an angle, such as at sunrise and sunset, is reflected even more because the amount of light reflected increases with the angle at which the light strikes the surface
b. Light penetration of a body of water drops exponentially with water depth as a result of absorption by water molecules and by suspended and dissolved substances; thus, light intensity diminishes rapidly with depth and greatly limits the presence of organisms that depend on light for survival
c. As light passes to greater depths, the water column absorbs some wavelengths of light more than others, so that the last remaining wavelength (color of light) penetrating to great depths is green; this color is ineffective for photosynthesis, consequently limiting the depth at which photosynthetic organisms can survive

## C. Seasonality
1. *Photoperiodism* is the observed response of an organism to the changes in length and timing of light periods
2. The daily and seasonal changes in the pattern of daylight are driven by the daily rotation of the earth on its axis and its annual revolution around the sun
3. These daily and seasonal changes produce a rhythmicity in plants and animals
4. Plants and animals display *circadian rhythms,* their basic activity patterns, which show a periodicity of approximately 24 hours
   a. These circadian activities suggest that organisms are able to measure and keep a sense of time
   b. The *biological clock* is an internal, poorly understood mechanism for measuring time; it is the means by which plants and animals can coordinate their activities with changing photoperiods
   c. Many animals display a circadian pattern of activity (for example, foraging, searching for mates, defending territory) and inactivity (for example, sleep)
   d. Some plants exhibit circadian patterns; for example, the common sorrel extends its leaves during the day and folds them close to the stem at night
5. Changes in periodicity from one season to another are known as seasonality
   a. One way plants display seasonality is in the time of flowering, which is largely a response to changes in day length
   b. Animals display seasonal changes or patterns that trigger reproductive cycles and migratory cycles

# VII. Nutrients

## A. General information
1. *Nutrients* are the materials taken in by organisms and used for growth and maintenance
2. Nutrients needed in relatively large amounts are known as *macronutrients;* others needed in small quantities are called *micronutrients*
3. Some nutrients are needed by all organisms; others are essential to specific species

## B. Nutrients and plants
1. Nutrient availability has an impact on the local and regional distribution of plants

2. Plants require some 16 essential nutrients, but individual species differ in the specific amounts and combinations of these nutrients that they require
3. Each species has a specific ability to exploit the nutrient supply that may not be identical to other species in the area; for example, two species growing in the same area may exploit slightly different nutrient sources because one species has a shallow root system and extracts nutrients from the upper layers of the soil, whereas the other species has deeper roots to tap nutrients at lower levels

## C. Nutrients and animals
1. All animals depend on plants for food, either directly (by consuming plant tissue) or indirectly (by eating other animals that consume plant tissue)
2. Nutrition often is more problematic for herbivores than for carnivores
   a. Herbivores' survival success often is controlled by food quality rather than quantity; there often is sufficient plant tissue available to allay hunger, but it may be of poor quality and thus affect reproduction, health, and longevity
   b. In most carnivores, food quantity is more important than quality
      (1) Carnivores rarely have dietary problems because they consume other animals that have resynthesized and stored protein and other nutrients from plants in their tissues
      (2) If they fail to obtain nutrients from an animal source, carnivores often resort to plant concentrates such as fruits

## D. Nutrient budgets
1. A *nutrient budget* is a measure of the input and outflow of elements (nutrients) through the various components of an ecosystem
2. Extra nutrients may enter an ecosystem through rainfall, stream flow, and animal migration, immigration, and emigration
3. Loss of materials and nutrients from the ecosystem can result from biological processes (for example, migration) and abiotic processes (for example, erosion, leaching, water flow)

---

# Study Activities

1. List several abiotic and biotic components of an ecosystem.
2. Using temperature as the factor, explain the concept of tolerance range.
3. State Shelford's law of tolerance, and explain some of its consequences for the distribution of plants and animals.
4. Use a sketch to show the global patterns of wind and how they circulate.
5. Construct a chart to describe the differences among ectotherms, endotherms, and heterotherms.
6. Draw a diagram to show the major ways that organisms gain heat from and lose heat to their surroundings.
7. Describe how light and photoperiod influence the activity of plants and animals.
8. Explain the concepts of shade tolerance and shade intolerance in trees.

# 3

# Soils

## Objectives

After studying this chapter, the reader should be able to:
- Describe the major physical characteristics of soil and their importance to natural ecosystems.
- List and describe the major soil horizon types.
- Explain the significance of such soil characteristics as pH, texture, color, and moisture conditions.
- Explain the processes that lead to the formation and development of soils.
- Describe the ten major soil orders, as well as how they are formed and where they are found.

## I. Soil Components

### A. General information
1. Soil is a complex mixture of tiny particles of inorganic matter (not derived from living organisms) and organic matter (derived from living organisms), water, air, and living organisms
2. Typically, soil is approximately 45% inorganic mineral material, 5% organic material, and 50% open space between the soil particles
3. The open (or pore) spaces between soil particles serve as air passages for carbon dioxide, oxygen, and water, as well as passageways through which water moves

### B. Soil layers
1. Most undisturbed soils consist of a series of distinctive horizontal layers (soil horizons), collectively known as the *soil profile*
2. Scientists consider up to a total of six horizons within a soil, but most soils have three or four of the major horizons (see *Soil Profile,* page 24)
   a. The O horizon, the uppermost layer, contains dead and decaying plant and animal tissue
   b. The A horizon, the topsoil, consists of partially decomposed organic matter, the bulk of plant roots, and some inorganic minerals
   c. The E horizon is the zone of **leaching,** the area through which dissolved materials move downward through the soil carried by water

**24** Soils

## Soil Profile

A soil profile consists of a series of layers arranged one atop the other. Each layer is referred to as a horizon and is designated by a letter.

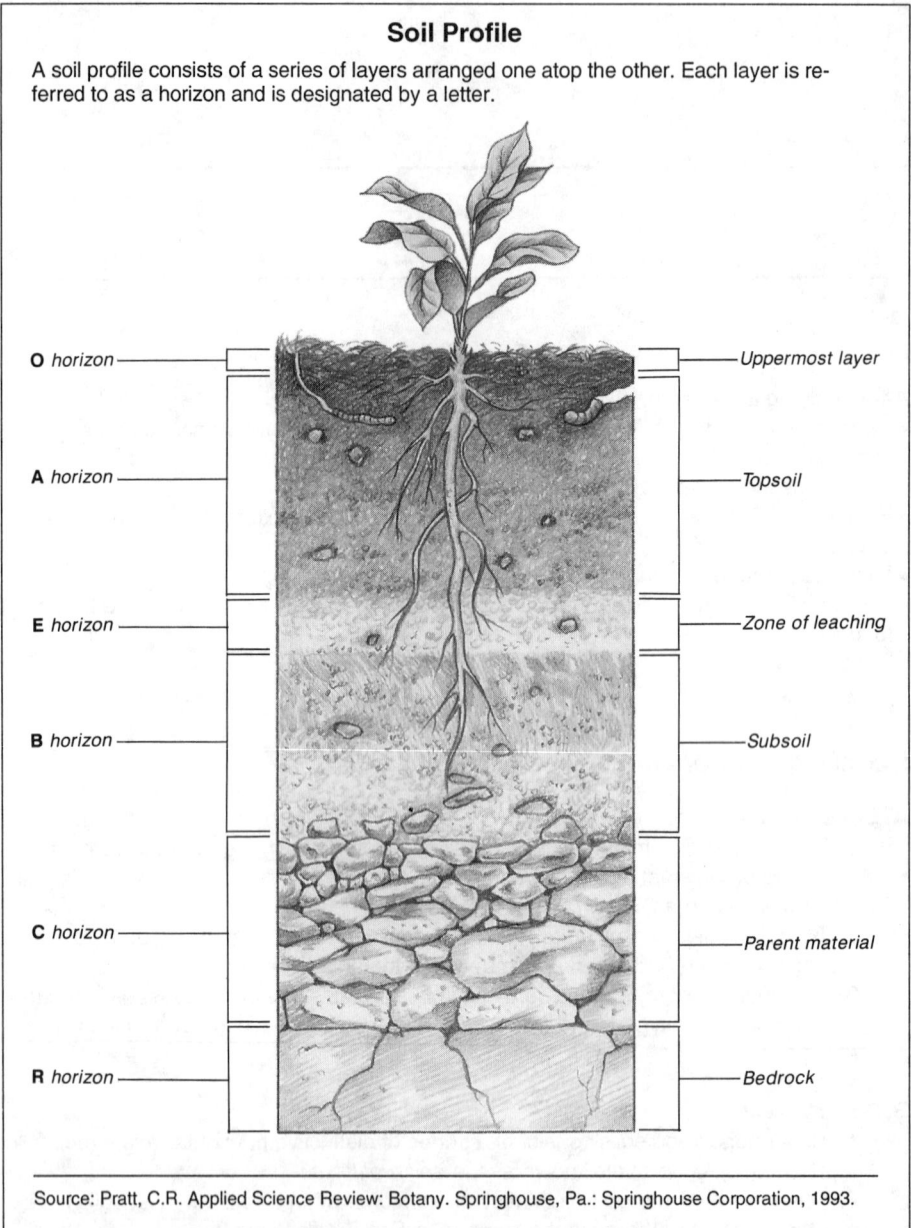

O horizon — Uppermost layer
A horizon — Topsoil
E horizon — Zone of leaching
B horizon — Subsoil
C horizon — Parent material
R horizon — Bedrock

Source: Pratt, C.R. Applied Science Review: Botany. Springhouse, Pa.: Springhouse Corporation, 1993.

    d. The B horizon, the subsoil, is the zone of *deposition,* where iron, aluminum, organic compounds, and clay accumulate from upper layers; it may have distinctive colors because of deposition

    e. The C horizon is the parent material from which the above layers are formed through weathering; it consists of partially broken-down inorganic materials

    f. The R horizon, the bedrock, is (except for fractures) an impenetrable layer of rock

3. Soil horizons demonstrate the decreasing influence of climate and the increasing influence of bedrock with increasing depth beneath the soil surface

## II. Soil Properties

### A. General information
1. There is a tremendous variety to soils; scientists classify approximately 15,000 different soil types
2. Differences in soil characteristics result from the type of parent rock, the mix of organic matter, the amount of water, the texture and age of the soil materials, vegetation, and a number of chemical factors

### B. Physical properties of soil
1. Soil *texture* is a measure of the size of the soil's individual mineral particles and the proportion of different-size particles
    a. Clay particles do not exceed 0.002 mm in diameter; silt particles range between 0.002 and 0.02 mm in diameter; sand particles range between 0.02 and 2 mm in diameter
    b. Texture has a major influence on the movement of air and water through a soil and on the penetration of roots into the soil
    c. Texture also influences the way soil particles clump together into larger lumps and clods
        (1) Soils with high clay content hold together strongly and, when wet, can form massive clumps
        (2) Soils with low clay and high sand content do not clump together
    d. Based on the proportion of clay, silt, and sand particles, soils are divided into textural classes
        (1) *Sandy soils* have a large proportion of sand particles
        (2) *Loams* are soils with approximately equal proportions of sand, silt, and clay particles
        (3) *Clay soils* have high proportion of clay particles
        (4) *Loam soils* have a crumbly, spongy quality that allows proper water drainage and air circulation; these generally are considered good agricultural soils
2. Soil color is an important characteristic for soil classification, though it is of relatively little importance to soil function
    a. Dark-colored soils found in temperate regions generally are higher in organic matter than lighter-colored soils
    b. Red- and yellow-colored soils generally indicate the presence of iron oxides
    c. Yellow-brown or gray colors often are indicative of poorly drained soils
    d. Soil colors are classified using standardized color charts determined by the U.S. Soil Conservation Service

### C. Chemical properties of soil
1. Soil chemical properties are determined primarily by oxygen, water, mineral, and organic matter content; available plant nutrients; and pH
2. Chemical elements required by plants in relatively large amounts for healthy growth and reproduction are called *macronutrients;* these include carbon, hy-

drogen, oxygen, nitrogen, phosphorus, potassium, calcium, magnesium, and sulfur
   3. *Micronutrients,* sometimes called trace elements, are required at very low levels; these include iron, manganese, molybdenum, zinc, copper, chlorine, and boron
   4. **Organic matter** consists of dead leaves, stems, and roots along with insect remains, animal droppings, and worm secretions, all of which accumulate in the upper portion of the soils
      a. Soils contain 1% to 7% organic matter, which serves as a home and source of food for microorganisms and acts as a sponge that soaks up and retains moisture
      b. Decomposer organisms (for example, bacteria and fungi) break soil organic compounds into simpler forms such as nitrate, phosphate, potassium, and sulfate, which are usable by plants
   5. *pH* is a measure of the hydrogen ion concentration in an aqueous solution
      a. Low pH (between 0 and 7) indicates a large number of hydrogen ions in a solution and is considered acidic
      b. A pH of 7 indicates a neutral solution
      c. High pH (between 7 and 14) indicates a smaller number of hydrogen ions and is considered alkaline
      d. A major source of soil acidity is the hydrogen ions formed when carbon dioxide in the atmosphere reacts with water in the soil to produce carbonic acid, which in turn dissociates to release hydrogen ions
      e. pH affects leaching (loss of nutrients) from soils and the availability of plant nutrients
      f. Most important plant nutrients are readily available at a pH of approximately 7

## D. Moisture
   1. A soil's *water-holding capacity* (the amount of water held by a soil) depends primarily on its porosity, or *pore space* (the amount of pores or open space in the soil), and on the size and relative proportion of soil particles
   2. If a soil is too porous, it will not hold water; if it is not sufficiently porous, it will retain water and become water-logged
   3. Because the total surface area of particles in the soil increases as particle size decreases, silty (high clay content) soils hold more water than coarse (sandy) soils
   4. Water-holding capacity is not completely equivalent to *water availability* (the amount of water available for use by plants); plant roots can easily extract water that clings loosely to soil particles but cannot extract water molecules bound tightly to soil particles
   5. The maximum amount of water that a soil can hold after all excess water has drained off is known as *field capacity;* field capacity is greatly influenced by soil texture and the relative proportions of clay, silt, and sand particles

## E. Air
   1. About 50% of a typical soil's volume is made up of pore spaces
   2. These spaces are filled with water in a waterlogged soil and with air in drier soils
   3. Normal, healthy plant growth requires well-aerated soils
   4. Oxygen is needed for respiration of roots and soil organisms and must diffuse into the soil's pore spaces from the atmosphere

5. Carbon dioxide is given off by the plant roots in concentrations reaching 10%, compared with the 0.03% concentration found in the atmosphere; it must diffuse out of soil pore spaces into the atmosphere to ensure a proper environment for healthy root growth

**F. Living organisms**
   1. Living organisms typically found in natural soils include bacteria, fungi, molds, protozoa, mites, nematodes, earthworms, insects, and burrowing animals
   2. These organisms make up only about 0.1% of the mass of soil but are important because they contribute to soil fertility and porosity and churn and mix the soil

## III. Soil Formation and Development

**A. General information**
   1. The formation and development of soil begins with the physical and chemical breakdown of rocks and their minerals
   2. Over time, the effects of bacteria, plants, and animals on soil development in a particular area become more pronounced, greatly altering soil characteristics, such as pH, nutrients, and texture

**B. Factors affecting soil development**
   1. *Weathering* is the decomposition of rock near the bottom of the soil profile or at the surface of newly exposed rock; it is promoted by both physical and chemical processes
      a. Physical weathering (or mechanical breakdown) involves the breaking down of parent rock into bits and pieces by exposure to temperature changes and the physical action of moving ice and water, growing plant roots, and human activities
      b. Chemical weathering involves chemical attack and dissolution of parent rock by exposure to rain water, surface water, and gases
   2. Climate (temperature, wind, rain, and ice) plays a major role in the rate of weathering of rock materials and thus influences soil development
      a. Temperature alters the rate at which chemical weathering takes place
      b. Wind, wave action, and glacial movement cause rock particles to move and rub against one another; this process of abrasion contributes to physical breakdown of parent rock
   3. As rock material is pulverized and weathered by physical and chemical activities, plants and animals begin to inhabit the developing soil
      a. Growing roots can exert enough pressure to split rock, thus facilitating weathering and soil development
      b. Lichens and mosses grow and penetrate between mineral grains and loosen rock particles
      c. Plants tend to trap windblown materials and augment soil development
      d. Plants also provide shade (thus altering moisture conditions), serve as windbreaks, and provide roots to anchor the developing soil and protect against erosion; in short, the development of soil and vegetation go hand in hand
      e. As organisms die and decompose, their bodies add organic material to the soil profile

**28** Soils

    4. Because erosion can play an important role in determining the rate at which a soil accumulates on a particular site, topography and slope have a major impact on the type of soil that ultimately develops in a region

## C. Soil formation
1. The categories of soil formation are based on the characterization of major physical, biological, and chemical factors that lead to the formation of a particular soil type in a particular area
2. Five major processes are involved in soil development or formation: podzolization, laterization, calcification, gleyization, and invertization
3. **Podzolization,** or spodsolization, occurs in moist climates with seasonal temperature variation, such as eastern North America
   a. In podzolization, iron and aluminum are leached from the A and E soil horizons and are deposited and concentrated in the B horizon
   b. In these temperate climates, winter temperatures slow decomposition of organic materials in surface and root zones
   c. Typically, in podzolic soils the A and E horizons are coarser than the B horizon, which has a claylike texture
   d. Typical vegetation in podzolic soils are deciduous and coniferous forests
4. **Laterization,** or oxisolation, occurs in warm moist climates, such as humid subtropical and tropical forested regions where wetting and drying of soil promotes oxidation of soil materials
   a. Weathering in these regions generally is chemical because of the uniformly high temperatures
   b. In laterization, iron and aluminum are oxidized to rock-hard and brick-red nodules in the B horizon; the resulting soil is deep red because of the presence of iron oxides
   c. Sodium, calcium, magnesium, potassium, and silicon move out of reach of plant roots and into ground water; this leaching process leaves an abundance of hydrogen ions behind, making soils formed by laterization very acidic
   d. The tropical rain forests of Mexico and Hawaii are the dominant vegetation on soils formed by laterization in North America
5. **Calcification** occurs in semiarid to arid regions where evaporation from the soils exceeds precipitation
   a. In calcification, calcium, sodium, magnesium, and potassium ions are drawn to and accumulate near the top of the soil profile because of the evaporation of water from the soil surface
   b. As a result of the repeated drawing of salts toward the soil surface, calcium carbonate ($CaCO_3$) nodules often accumulate in the B horizon
   c. Calcified soils usually are neutral to slightly basic (pH 6.5 to 8.0) and generally are very productive when irrigated
   d. Plains, prairies, and desert regions are typical areas where calcification occurs
   e. In very dry climates, salts accumulate in greater concentration near the surface and result in an extreme calcification condition known as **salinization,** which often occurs as a result of prolonged irrigation of soils in arid and semiarid areas

(1) Agriculturally valuable soils are destroyed by salt accumulation because of the addition of irrigation water, which draws salts from lower soil layers to the surface as it evaporates
(2) Cropland with severely salinized soil must be taken out of production for 2 to 5 years; underground drainage pipes must be installed and the soil flushed with large quantities of water to remove the accumulated salts
(3) Approximately 18% of the world's cropland currently is irrigated and produces about one-third of the world's food; it is projected that by the year 2020, the percentage of irrigated land will have at least doubled, placing a large acreage of land in danger of salinization
(4) Salinization cannot be stopped, but it can be slowed by periodic flooding of the soil as necessary and by the use of more efficient irrigation practices that reduce evaporation of water from the soil surface
(5) Waterlogging is a common problem accompanying soil salinization in dry regions
    (a) To combat the problems of salinization on irrigated land and prolong the useful life of the soil, farmers add more and more water to flush away salts or to leach them deeper into the soil
    (b) The additional water causes the water table to rise and, as a result, salt-laden water begins to surround plant roots and kill crops
    (c) It is estimated that at least one-tenth of all irrigated land currently suffers from waterlogging and subsequent decline in agricultural production
6. **Gleyization** occurs in areas of poor drainage or standing water (water accumulation in the B or C soil horizons for long periods of the year)
   a. Because of the poor drainage, the soils typically are waterlogged and therefore low in oxygen, which in turn results in no oxidation of soil materials
   b. A sticky, blue-gray accumulation of clay and **humus** forms in the B horizon and gives these soils their characteristic coloration
   c. The breakdown of organic materials often is incomplete, and these materials may accumulate as an organic layer (peat) in the O horizon
   d. As microorganisms convert vegetation to humus, the soil water becomes acidic
   e. Gleyization occurs in coastal wetlands and in the permafrost regions of the northern tundra climates
7. **Invertization** occurs in climates where there are distinct wet seasons followed by periods of extended evaporation and drying
   a. Invertization results in an inverted soil profile and occurs in soils with a high clay content near the surface
   b. The invertization process requires substantial water from precipitation followed by prolonged dry periods, during which the upper-surface clay begins to crack; this allows some of the O and A horizon material to fall into deep fissures, which can extend into the B horizon
   c. Repeated drying and cracking results in a soil that appears to have the A horizon below the B horizon
   d. Invertization occurs in prairie and grassland areas

## IV. Soil Classification

### A. General information
1. Soils are sometimes grouped according to their origin or location
   a. *Residual soils* form and develop where they are found, from bare rock exposed to physical and chemical weathering
   b. *Transported soils* are those transferred from one area to another as loose sediments by wind, water, landslides, or human activity
2. The U.S. Comprehensive Soil Classification System, which consists of 10 soil orders, was developed to standardize soil nomenclature

### B. U.S. Comprehensive Soil Classification System
1. *Alfisol* soils are formed by podzolization, where humus is restricted to the upper surface; they have well-developed horizons and typically are the gray-brown soils of wooded areas
2. *Aridisol* soils have a low humus content, are dry for extended periods, and may display some calcification; they typically are desert soils
3. *Entisol* soils do not have distinct horizons and may be subject to erosion; they typically are the soils of **alluvial** regions, commonly found on floodplains
4. *Histosol* soils generally are formed by gleyization and have a high organic content; they commonly are found in bogs
5. *Inceptisol* soils are formed by podzolization with significant erosion; they have a fine texture and often a shallow profile
6. *Mollisol* soils result from podzolization and some calcification; they typically are dark and rich in nutrients and are found in semihumid prairie regions
7. *Oxisol* soils result from laterization; they are highly weathered and often red, yellow, or gray, found in the tropic and subtropical regions (for example, Florida, Hawaii, and Mexico)
8. *Spodosol* soils are formed by podzolization; they are light gray and usually are high in iron and aluminum concentration
9. *Ultisol* soils, also formed by podzolization, are intensely leached and weathered soils containing a large proportion of clay; they generally are found in warm climates of the southeastern United States
10. *Vertisol* soils, formed by invertization, are dark clay soils that exhibit wide, deep cracks when dry; they commonly are found in Texas and similar regions

---

## Study Activities
1. Compare and contrast the physical and chemical weathering processes in soil formation.
2. Sketch a soil profile; label the major regions, and briefly characterize them.
3. Summarize the U.S. Comprehensive Soil Classification System. What are the major soil orders? How are they determined?
4. Construct a table of soil formation processes. List the five processes in the first column, a brief description in the second column, and the types (soil orders) of soils produced by each process in the third column.
5. Define these soil moisture terms: field capacity and water availability.
6. Compare and contrast residual soils and transported soils with respect to how and where they are formed.

# 4

# Energy Flow

## Objectives

After studying this chapter, the reader should be able to:
- Distinguish among primary production, secondary production, and productivity.
- Explain the concepts of food chain and food web.
- Describe detrital and grazing food chains, and explain their importance in ecosystems.
- Explain the importance of ecological pyramids of number, biomass, and energy.
- Describe the significance and importance of ecological efficiency.

## I. Biomass, Productivity, and Production

### A. General information
1. The sun bombards the earth with solar radiation in two forms important to the biosphere
   a. One important form of energy is heat, which warms the earth, drives the water cycle, and generates the water and air currents of the atmosphere and the oceans
   b. The other is that portion of the solar radiation spectrum used in photosynthesis to produce carbohydrates and other compounds important to organisms
2. The flow of energy through an ecosystem begins with photosynthesis and continues as one group of organisms consumes another
3. The amount of energy accumulated by an organism is referred to as **production**
   a. Production, or energy accumulation and storage, usually is expressed in mass per square meter per year (for example, as kilocalories per square meter per year [kcal/m$^2$/year]); it also may be expressed as organic matter accumulated per year (for example, as grams per square meter per year [g/m$^2$/year])
   b. Production generally is classified as primary production when dealing with plants and secondary production when considering consumer organisms
   c. The rate of production is referred to as **productivity**
4. As organisms accumulate energy, they usually convert it to tissue; this accumulated tissue is referred to as biomass

### B. Biomass
1. The tissue of organisms that accumulates over time is known as *biomass,* or the total weight of living tissue

2. At any one time, each group of organisms contains some amount of energy stored as biomass — termed the *standing crop*
3. In temperate areas, a portion of the biomass accumulated is recycled seasonally as leaves drop and then decompose
4. Biomass usually is measured as grams of dry weight of organic matter per unit area, typically per square meter ($g/m^2$)
5. Biomass is an important parameter to measure because it indicates how much of an organism is present; if biomass is measured repeatedly over time, changes can reveal whether an organism is increasing or decreasing

## C. Primary production
1. *Primary production* is the energy accumulated by photosynthetic plants and is the first and most basic form of energy storage in an ecosystem
2. Primary production of ecosystems is influenced by temperature and precipitation because it is based on photosynthesis, which is influenced by these environmental factors
3. Primary production occurs in two forms: gross primary production and net primary production
   a. *Gross primary production (GPP)* is a measure of the total of all the sun's energy that is assimilated; therefore, GPP is equivalent to total photosynthesis
   b. *Net primary production (NPP)* is the energy remaining as stored organic matter after energy for reproduction and maintenance is deducted
      (1) The energy for reproduction and maintenance is derived from respiration, so these costs often are referred to as *respiration costs*
      (2) NPP is the energy available to the next group of organisms that consume plant tissue
      (3) Theoretically, all of NPP is available to the consumers in an ecosystem; however, much of it is either left unused or is transported from the ecosystem by humans, wind, or water
      (4) Annual NPP is influenced by the age of the ecosystem and changes with time and age
4. Ecosystems with a large biomass have large metabolic costs (respiration rates) and, therefore, low ratios of net to gross primary productivity
5. The rate at which energy is stored by plants is considered *primary productivity* and determined by measuring the energy accumulated per unit of time
6. Wetland (swamps, marshes) and estuary regions are among the most productive ecosystems, whereas deserts and open ocean areas have the lowest net productivities

## D. Secondary production
1. *Secondary production* is the accumulation of energy by consumer organisms (animals and decomposers such as bacteria and fungi)
2. The conversion of primary production to secondary production first occurs when plant tissue is consumed by herbivores
3. The quantity of net production harvested by herbivores is influenced by the type of herbivore, its feeding and digestive processes, and the density of the herbivore population
4. Once consumed, a considerable amount of the plant material ingested may pass through the animal's body unassimilated as waste

- a. A grasshopper (poikilothermic) assimilates about 30% of the grass it consumes
- b. A mouse (homoiothermic) assimilates about 85% to 90% of what it consumes
5. Patterns of consumption also vary from organism to organism and at different times, so that some net production will be left untouched or as "leftovers"
6. Once assimilated, the energy content of the plant tissue either is diverted to maintenance and respiration, growth, and reproduction or is eliminated as waste material (feces or urine)
7. Energy left over after maintenance and respiration costs are met is used to fuel the production of new tissue, fat tissue, growth, and new individuals (reproduction); this production is available to other consumers that may feed on the herbivore
8. Secondary production is limited by the quantity and quality of net primary production and by the metabolic characteristics of the consumer
    - a. Homoiotherms generally have high assimilation rates (that is, they assimilate much of what they consume); they also have high metabolic costs, using about 98% of that energy in metabolism, with only 2% remaining as secondary production
    - b. Poikilotherms generally are not as efficient as homoiotherms at assimilation; they have lower metabolic rates and use about 56% of their total assimilation in metabolism, converting approximately 44% of the assimilated energy to secondary production
    - c. Poikilotherms are approximately 30% efficient at assimilation, whereas homoiotherms are approximately 70% efficient; therefore, poikilotherms must consume more calories than homoiotherms to obtain sufficient energy to meet their energy needs

## II. Food Chains and Webs

### A. General information
1. The energy stored by plants is passed along through the ecosystem as one organism consumes another in a series of steps known as a **food chain**
2. Because the food chains within an ecosystem can be complex and have many cross links (that is, more than one organism consuming a particular food item or one organism consuming more than one organism as food), the food chains often are referred to as **food webs**

### B. Food chain components
1. Organisms in an ecosystem can be categorized as either producers (plants) or consumers
2. Consumers generally are classified as herbivores, carnivores, omnivores, or decomposers; other feeding groups include **parasites** and *scavengers* (herbivores and carnivores that eat only dead material)
3. Primary producers form the basis for all food chains and food webs, because photosynthetic plants capture sunlight
4. As one group of organisms consumes the next, energy passes along the food chain; each group of organisms can be considered a temporary stopping point for the energy of the ecosystem (see *A Typical Food Web,* page 34)

34   Energy Flow

## A Typical Food Web

In this diagram of a food web, the arrows indicate the different pathways by which energy passes through the ecosystem as one organism is consumed by another. This same diagram can be simplified or condensed to emphasize the basic trophic or feeding structure of the ecosystem, as shown to the right of the diagram.

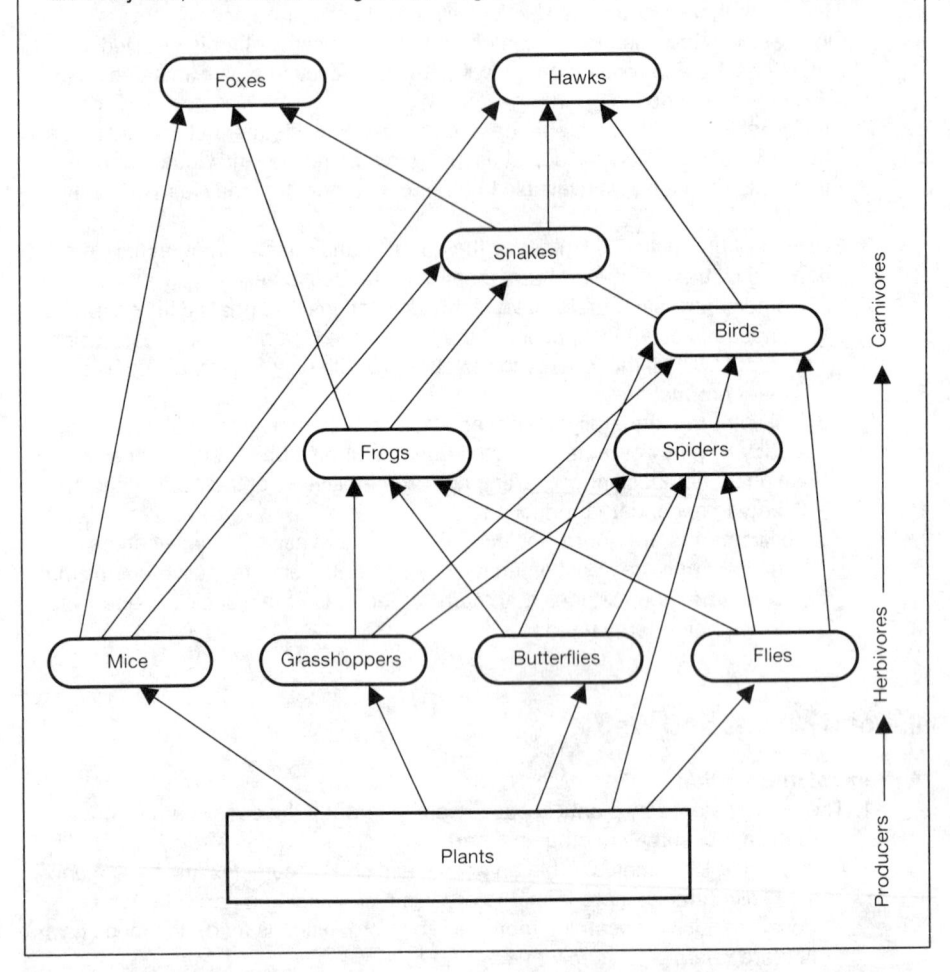

5. The levels of a food chain consist of various groups of organisms (for example, herbivores, carnivores, decomposers) and are referred to as ***trophic levels,*** or feeding levels

## C. Major food chains
1. In any ecosystem, there are two major food chains: detrital and grazing
2. The *detrital food chain* is the dominant food chain of most terrestrial and shallow-water ecosystems; as such, it is the major pathway of energy flow in these systems
   a. In ecosystems where the detrital food chain predominates, little of the net productivity is consumed by grazing herbivores, and the bulk of the net pro-

duction is utilized by **detritivores,** which feed on partly decomposed plant and animal tissue
    b. The percentage of production that goes into the detrital food chain in some ecosystems can be substantial and is best appreciated by examining the following examples:
        (1) In a tulip poplar *(Liriodendron)* forest, 50% of the GPP goes into maintenance and respiration, 13% is accumulated as new plant tissue, 2% is consumed by herbivores, and 35% goes into the detrital food chain
        (2) In a typical deciduous forest, 55% of GPP goes into plant respiration, 3% to herbivores, 20% into biomass accumulation, and 22% into the detrital food chain
3. The *grazing food chain* is conspicuous and obvious in terrestrial ecosystems; it consists of herbivores (for example, cattle, rabbits, insects, gazelles, deer) consuming primary production
    a. In terrestrial ecosystems, only a very small percentage of the GPP passes through the grazing food chain
    b. For example, investigators found that even heavily grazing cattle consume only 15% of the total net aboveground production in a grassland; furthermore, 40% to 50% of the energy consumed by the cattle is returned to the ecosystem via the detrital food chain as feces
    c. Although aboveground herbivores are its most conspicuous consumers, the grazing food chain includes consumers of underground plant parts (that is, insects, nematodes, and other soil biota); this underground consumption can have a major impact on primary production and the grazing food chain
4. In general, as energy passes from one level to the next, there is a reduction in available energy
    a. The energy is reduced in magnitude by a factor of 10 from one level to the next; thus, if 10,000 kcal of plant energy is consumed by herbivores, approximately 1,000 kcal is converted to herbivore tissue; if these herbivores are consumed by other animals, approximately 100 kcal is converted into their flesh
    b. As a result of this energy loss at each transfer, the length of food chains is limited; most food chains have three or four trophic levels

## D. Trophic levels and ecological pyramids
1. An ecological pyramid is a graphical representation of the biomass, census numbers, or energy content of the various trophic levels of an ecosystem (see *Ecological Pyramids,* page 36)
2. Each trophic level is represented as a layer or segment in the pyramid
    a. The first trophic level consists of producers; the second trophic level, of herbivores; and the higher trophic levels, of carnivores
    b. Some animals occupy a single trophic level; others occupy more than one level
3. A *pyramid of numbers* is constructed by conducting a census of each trophic level of an ecosystem
    a. The organisms at the lower end of the food chain are the most numerous
    b. Organisms at successive levels decrease rapidly in number until there are only a few carnivores at the top level, resulting in a generalized pyramid shape

## Ecological Pyramids

Graphical representation of chains may be accomplished by examining the number of individuals in each trophic level (A), the energy content of each trophic level (B), or the biomass of living tissue in each trophic level (C). Some ecosystems possess primary producers with a very small biomass but a very high productivity, which can result in an inverted pyramid of biomass (D). The numbers used here are for illustrative purposes only.

**(A) Pyramid of numbers**

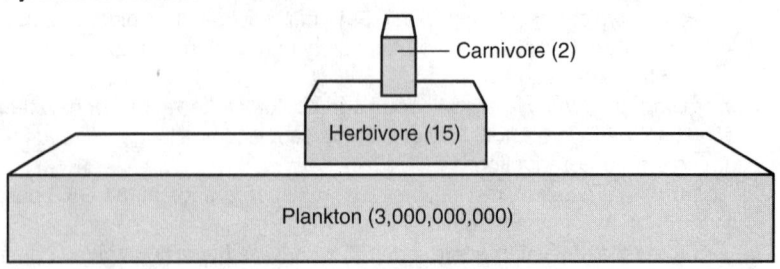

- Carnivore (2)
- Herbivore (15)
- Plankton (3,000,000,000)

**(B) Pyramid of energy**

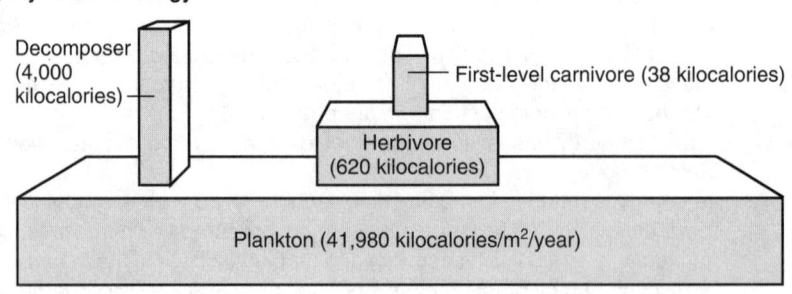

- Decomposer (4,000 kilocalories)
- First-level carnivore (38 kilocalories)
- Herbivore (620 kilocalories)
- Plankton (41,980 kilocalories/m$^2$/year)

**(C) Pyramid of biomass**

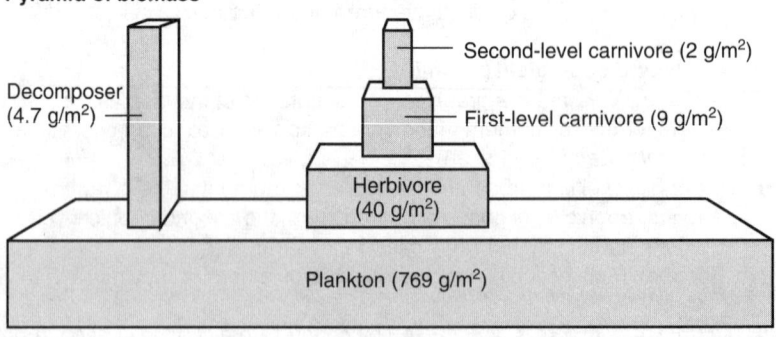

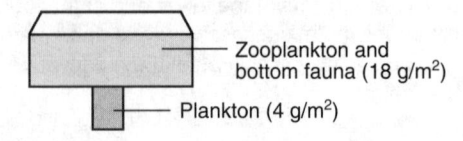

- Second-level carnivore (2 g/m$^2$)
- Decomposer (4.7 g/m$^2$)
- First-level carnivore (9 g/m$^2$)
- Herbivore (40 g/m$^2$)
- Plankton (769 g/m$^2$)

**(D) Inverted pyramid of biomass**

- Zooplankton and bottom fauna (18 g/m$^2$)
- Plankton (4 g/m$^2$)

4. By summing the living biomass of organisms in each trophic level of an ecosystem, a *pyramid of biomass* can be constructed
   a. A pyramid of biomass indicates by weight a measurement of the energy present at a given time
   b. Because energy or material is lost in each successive transfer in a food chain, the total mass supported at each trophic level is limited by the rate at which energy is stored at the next-lower trophic level
   c. Characteristically, the biomass of producers is greater than the biomass of the herbivores they support; in turn, the biomass of herbivores is greater than that of the carnivores they support
   d. In some ecosystems, such as very productive aquatic systems (for example, lakes and ponds) in which the producers are represented by microscopic algae, the pyramid of biomass may be inverted
      (1) The producer algae of these ecosystems have characteristically short life cycles and reproduce rapidly, but the ecosystems accumulate little organic matter
      (2) Zooplankton are the predominate herbivores in these ecosystems
      (3) Because the producer algae are very small, their biomass at any given time is small; as a result, the base of the pyramid is much smaller than the structure that it supports
5. A *pyramid of energy* is constructed by measuring productivity or rate of energy accumulation in each trophic level
   a. Because of the inefficiency of energy transfer from one trophic level to another when one organism consumes another, energy is lost from the food chain at each transfer; so, the energy content of each successive trophic level decreases
   b. A pyramid of energy always has a triangular or pyramidal shape because each higher trophic level contains less energy than the others preceding it
   c. Pyramids of energy are important because they demonstrate the role of various organisms in the transfer of energy within the ecosystem

## E. Ecological efficiency
1. *Ecological efficiency* is the percentage of biomass produced by one trophic level that is incorporated into the biomass of the next higher trophic level
2. A wide range of ecological efficiency exists among the various feeding groups that comprise an ecosystem; as a rule of thumb, approximately 10% of the energy in one trophic level is transferred to the next trophic level, or, conversely, approximately 90% of the energy in any one trophic level is unavailable to the next trophic level
3. Ecological efficiency limits the length of food chains because of loss of energy at each transfer

# Study Activities
1. Choose a simple ecosystem (for example, a pond) and sketch its food web structure. Explain where energy is lost and transferred from one trophic level to the next.
2. Explain the underlying cause for the shape of an ecological pyramid of energy.
3. Describe how it is possible for a pyramid of biomass to be inverted.
4. Explain why most food chains consist of a maximum of three or four trophic levels.

# 5

# Movement of Materials Through Ecosystems

## Objectives

After studying this chapter, the reader should be able to:
- List the major biogeochemical cycles.
- Explain the difference between gaseous and sedimentary cycles.
- Briefly describe the flow of oxygen, carbon, water, nitrogen, phosphorus, and sulfur through the biosphere.
- List the major compartments or features of each of the biogeochemical cycles discussed.
- Identify the major constituents of the global pollution problems of ozone shield depletion, global warming, and acid precipitation.
- Describe the phenomenon of biological magnification.

## I. Biogeochemical Cycles

### A. General information
1. Organisms rely for survival on the flow of energy and the circulation of materials (nutrients) through the ecosystem
2. The flow of energy and the circulation of materials are interconnected; the energy flow is essential for mineral cycling, and the availability of nutrients limits primary productivity, which in turn influences the energy flow through the rest of the food chain
3. The cyclic movement and recycling of materials through an ecosystem is termed a *biogeochemical cycle*
4. Biogeochemical cycles involve both abiotic and biotic components; atoms are incorporated into the complex molecules of living organisms, passed from one trophic level to the next in food chains, and, with death and decomposition of these organisms, are released to replenish nutrient pools that plants and other autotrophs use to synthesize new organic matter
5. A biogeochemical cycle consists of two large compartments or locations where materials are found: the reservoir compartment and the exchange, or cycling, compartment
   a. The *reservoir compartment,* the larger of the two compartments, usually is abiotic and contains materials largely unavailable to living organisms
   b. The smaller *exchange compartment* is more active and more readily available to organisms; it is subject to rapid exchanges between organisms and their immediate environment

6. There are two types of biogeochemical cycles: gaseous and sedimentary

**B. Gaseous cycles**
1. In *gaseous cycles,* the major reservoir is the atmosphere; thus, these cycles are global in scale
2. The gaseous cycles involve the major gases important to organisms and include oxygen, carbon, nitrogen, and water vapor
3. The *oxygen cycle* involves the global movement of oxygen between the atmosphere, oceans, sediments, and living organisms
    a. The oxygen cycle is linked to the cycling of carbon, nitrogen, and sulfur in the biosphere
    b. The atmosphere contains 21% oxygen, provided by two important sources: chemical breakdown of water in the upper atmosphere and photosynthesis
        (1) In the upper atmosphere, ionizing radiation from the sun (in the form of ultraviolet light) chemically degrades water vapor to release oxygen and hydrogen
        (2) Ultraviolet light splits water molecules to release hydrogen and oxygen; the hydrogen escapes into space and the oxygen becomes part of the atmospheric reservoir
        (3) Photosynthetic plants release oxygen into the atmosphere as a by-product of photosynthesis
    c. Oxygen is chemically very active; it can combine with many different chemicals to form various biologically important molecules
    d. Oxygen is cycled or exchanged within the biotic components of an ecosystem in the form of carbohydrates, proteins, nitrates, sulfates, lipids, and other molecules of living tissues
    e. Oxygen also reacts with organic matter and mineral materials in the soil and sediments beneath lakes, streams, and oceans
    f. The oxygen trapped in soil and sediments is released by the activity of certain sulfur-bacteria strains
    g. The atmospheric reservoir of oxygen is very large compared to the oxygen in movement through the various other compartments of the oxygen cycle
    h. In the upper atmosphere, the sun's ultraviolet radiation converts some of the oxygen ($O_2$) to ozone ($O_3$); this ozone forms a protective layer in the upper atmosphere known as the ozone layer
4. The *carbon cycle* involves the global movement of carbon within the biosphere (see *The Carbon Cycle,* page 40)
    a. Carbon is a basic component of organic compounds and a major element in the fixation of energy by photosynthesis and release of energy during respiration
    b. During photosynthesis, carbon dioxide ($CO_2$) is removed from the atmosphere (or, in the case of aquatic plants, from the water) and incorporated into the organic compounds of the plant body
    c. As the various organic compounds are used by the plant itself, some $CO_2$ is returned to the atmosphere, but much is retained within the plant body
    d. As plants are consumed by herbivores and the herbivores, in turn, are consumed by carnivores, the carbon compounds are passed from one trophic level to the next
    e. As waste is produced or as organisms die, decomposer organisms (bacteria and fungi) obtain carbon from the decaying tissues

## The Carbon Cycle

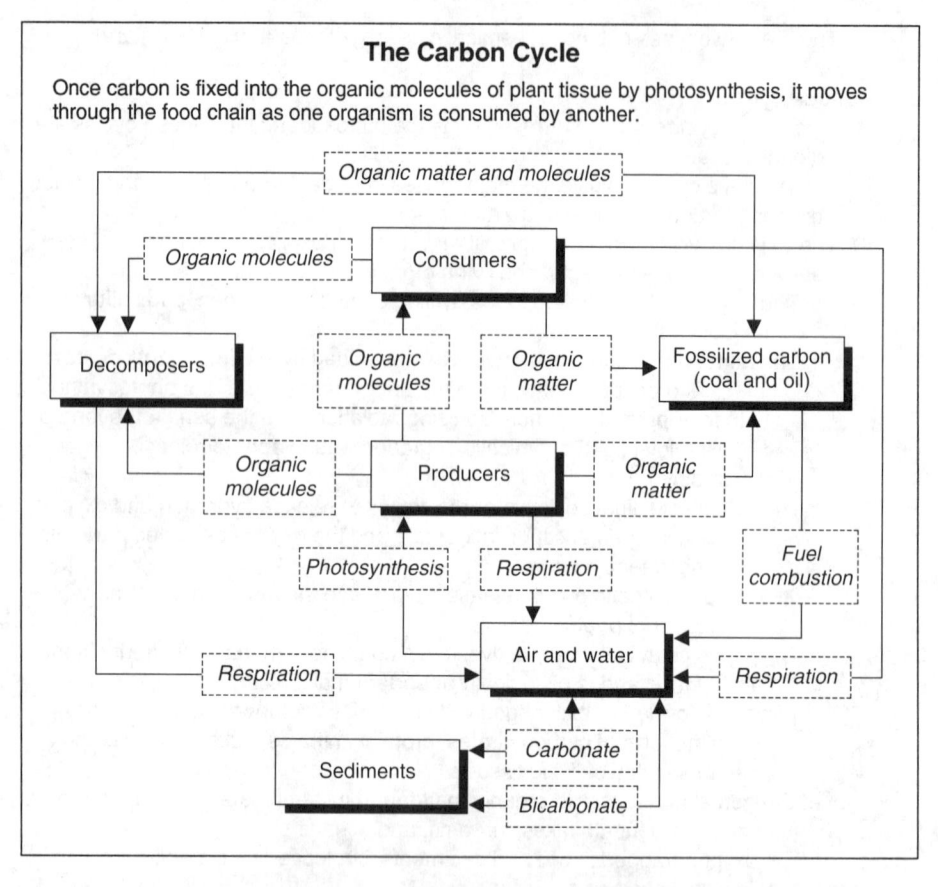

Once carbon is fixed into the organic molecules of plant tissue by photosynthesis, it moves through the food chain as one organism is consumed by another.

*[Diagram: Flow chart showing carbon cycle with boxes for Decomposers, Consumers, Producers, Fossilized carbon (coal and oil), Air and water, and Sediments. Arrows labeled Organic matter and molecules, Organic molecules, Organic matter, Photosynthesis, Respiration, Fuel combustion, Carbonate, and Bicarbonate connect these components.]*

  f. All consumer organisms (as well as decomposers) carry on respiration, which releases $CO_2$ into the atmosphere
  g. When $CO_2$ dissolves in water, some of it chemically reacts with the water to form carbonic acid, which may in turn react further to form compounds known as carbonates and bicarbonates
   (1) Some carbonate compounds are not very soluble in water and may precipitate to form deposits in sediments of lakes and oceans
   (2) Carbonate and bicarbonate ion formation are reversible reactions; they may proceed in either direction, forming carbonates (and bicarbonates) or resulting in the release of $CO_2$
   (3) Because these reactions are reversible, the effect is to buffer the $CO_2$ content of the atmosphere: if there is a local increase in carbon dioxide in the air, more dissolves in water to form carbonic acid; if there is a local decline in atmospheric carbon dioxide, the reactions are reversed and carbon dioxide is released into the air
   (4) Carbon incorporated into lake and ocean sediments probably returns to circulation very slowly; it may be brought back into circulation by geological events, such as volcanoes or the uplift of land and subsequent weathering of limestone deposits (calcium carbonate)
  h. The storage of fossil carbon in the form of fossil fuels is an additional aspect of the carbon cycle of great importance to humans

## The Nitrogen Cycle

The major reservoir for nitrogen is the atmosphere, where nitrogen is stored as a molecular gas; in this form, it is largely unavailable to organisms. Through the process of nitrogen fixation, nitrogen gas is converted to forms usable by living organisms and cycled through the ecosystem as amino acids, proteins, and nucleic acids.

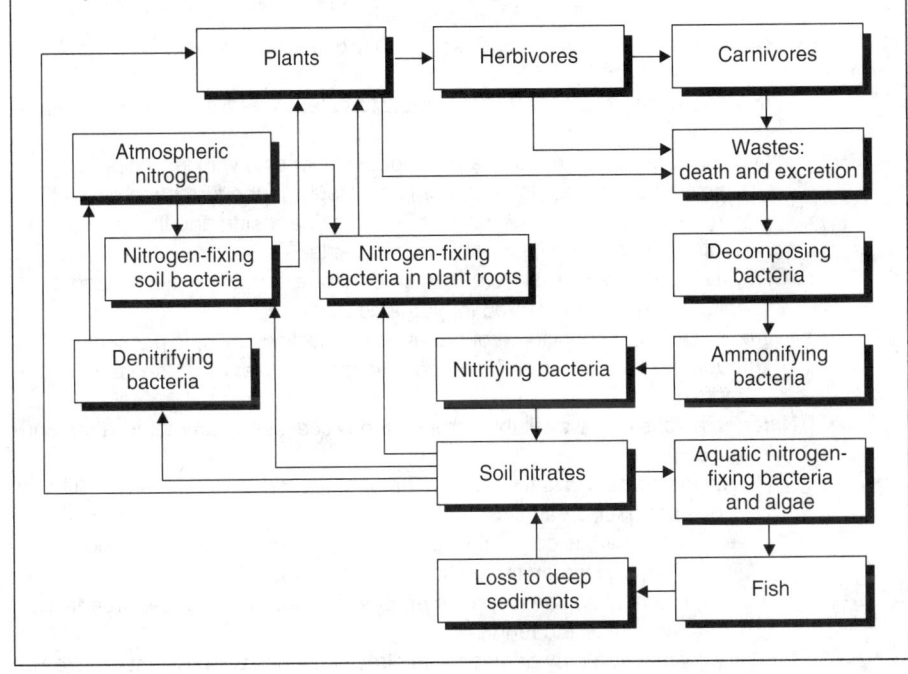

- (1) During the Carboniferous period (280 to 350 million years ago), a large amount of organic matter collected in ancient swamps and bogs; this material was converted to the *fossil fuels:* coal, gas, and petroleum
- (2) The fossil fuels are a storehouse of carbon that, under natural conditions, probably would decompose slowly or remain sequestered and out of circulation in the carbon cycle
- (3) Human exploitation of fossil fuels as an energy source through burning results in substantial input of carbon into the atmosphere
- (4) Much of the carbon released into the atmosphere is thought to have gone into the buffering system of the oceans; the rest remains in the atmosphere and may be contributing to global atmospheric changes

5. The *nitrogen cycle* involves the movement of nitrogen and nitrogen-containing compounds through the biosphere (see *The Nitrogen Cycle*)
    a. The atmosphere is a large pool or reservoir of nitrogen, most of which (79% of the atmosphere) is in the gaseous form ($N_2$), which is not available to most organisms
    b. $N_2$ is converted from its gaseous state to ammonia or nitrate in a process known as **nitrogen fixation;** this results from physical or biological processes
       (1) Physical processes that can fix nitrogen include lightning, volcanic activity, and meteors

(2) Most nitrogen fixation (approximately 90%) is biological fixation, which results from the activity of nitrogen-fixing bacteria and algae

c. Biological fixation may be performed by free-living organisms found in soil or water or by organisms living in specialized root structures of certain plants

  (1) The general chemical reaction of biological nitrogen fixation is:
  $$N_2 \rightarrow 2N \rightarrow 2N + 3H_2O \rightarrow 2NH_3 + \tfrac{3}{2}O_2$$
  (2) Free-living bacteria responsible for nitrogen fixation include *Azotobacter* and *Clostridium*
  (3) Blue-green algae (that is, cyanobacteria, such as *Nostoc* and *Anabaena*) also fix nitrogen
  (4) *Rhizobium* bacteria form a symbiotic relationship with legumes (for example, beans, clover, alfalfa) that results in the formation of swellings, or root nodules, in which the bacteria reside and fix atmospheric nitrogen, making it available to the plants

d. Once nitrogen is converted to ammonia or nitrate, it may be assimilated by plant roots and incorporated into organic matter

e. Nitrogen (a major constituent of proteins and nucleic acids) is passed through the food chain and incorporated into the tissues of the various consumers

f. Nitrogen also is made available through the processes of ammonification and nitrification

  (1) **Ammonification** is the conversion of amino acids to ammonia ($NH_3$) by decomposer organisms
      (a) Amino acids combine with oxygen to form ammonia, carbon dioxide, and water; in addition, energy is released
      (b) Ammonification results in energy yield for decomposer organisms (bacteria and fungi)
      (c) Ammonia may be absorbed directly and readily by plant roots and subsequently is used to synthesize amino acids and proteins
  (2) **Nitrification** is a biological process in which ammonia is oxidized (chemically combined with $O_2$) to form nitrite ($NO_2^-$) and nitrate ($NO_3^-$)
      (a) The nitrite bacteria *Nitrosomonas* uses ammonia and oxygen to form $NO_2^-$ and water
      (b) The nitrate bacteria *Nitrobacter* uses nitrite and oxygen to form $NO_3^-$
      (c) Nitrification generally is considered a beneficial process, converting nitrogen to a form readily available to plants
      (d) Nitrification may be detrimental in areas of abundant rainfall because large amounts of $NO_3^-$ (a highly soluble form of nitrogen) are produced and subsequently leached from soils; thus, nitrate is lost from the ecosystem and may end up in water supplies as a pollutant

g. **Denitrification** is the process by which organic forms of nitrogen are converted to $N_2$ and released into the atmosphere by denitrifying bacteria (for example, *Pseudomonas*), which consume nitrates to obtain oxygen

6. The *water* or *hydrologic cycle* involves the movement of water through the biosphere between the ocean, the atmosphere, and the earth's surface (see *The Hydrologic Cycle*)

## The Hydrologic Cycle

The movement of water through the biosphere is referred to as the hydrologic cycle. The atmosphere and oceans serve as major reservoirs of water on a global scale, and the sun is the major energy source providing heat to promote evaporation and circulation.

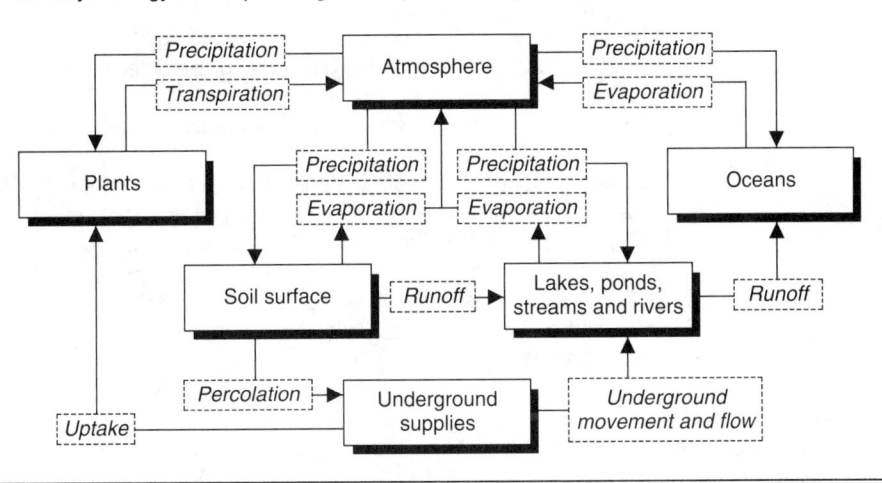

a. When radiant heat (solar energy) from the sun heats the surface waters of oceans, lakes, and rivers, the water evaporates and moves upward into the atmosphere, where it forms clouds
b. This evaporation process from the ocean surface separates the water molecules from the salty ocean waters and creates large amounts of pure water, which later becomes the freshwater resources found on land
c. When clouds of water vapor cool in the atmosphere, the water vapor condenses to form precipitation, which then falls either into the oceans or onto the land; in this way, precipitation is important in transferring water from the atmosphere to the ground surface or water body
d. As precipitation reaches the earth's surface, some of it is intercepted by vegetation, and some falls on rock surfaces and soil surfaces
e. Some of the environmental surfaces (bare ground, rock surfaces) are impervious, so the water evaporates directly back into the atmosphere
f. A portion of the water reaching the soil surface is absorbed and moves (infiltrates) into the ground; this portion is determined by slope, soil condition, soil type, vegetation cover, and magnitude and duration of the precipitation itself
g. When precipitation exceeds infiltration rate, the excess water tends to flow off the surface; this is referred to as *runoff*
   (1) In undisturbed forests, infiltration rates generally are greater than the intensity of precipitation; surface runoff is minimal because the combination of vegetation and thick soil layers slows the movement of the water and acts as a sponge
   (2) Bare, compacted soil (for example, a trampled lawn) has a low infiltration rate, and surface runoff can be substantial

      h. Once in the soil, water percolates through upper soil layers, eventually reaching the impervious rock or clay layer, where it accumulates, filling in the pores (spaces and cracks) in the soil and rocks and forming the *water table*
      i. Water that reaches the impervious layer gradually flows to springs or streams and eventually makes its way back to the oceans
      j. A portion of the water that enters the soil is retained in the soil
        (1) The portion of water held or retained between soil particles is called *capillary water*
        (2) A thin film of water molecules adhering to soil particles is called *hygroscopic water*
        (3) The maximum amount of water that a soil can hold (store) after all excess water has drained off is known as *field capacity* or *storage capacity*
          (a) Highly porous sandy soils have a low field capacity; fine-textured clay soils have a higher field capacity
          (b) Soils in urban areas often have been compacted and the soil texture disrupted by human activities such that their field capacity often is small

## C. Sedimentary cycles

1. In sedimentary cycles, the major reservoirs are the rocks and minerals of the earth's crust
2. Mineral elements required by organisms are obtained ultimately from inorganic sources — for example, salts dissolved in the water of soil, lakes, streams, and oceans
3. Mineral salts are released directly from the earth's crust by weathering
4. Plants and some animals obtain their mineral needs from mineral solutions in their environments (for example, plants obtain minerals from those dissolved in the soil water surrounding their roots); other animals obtain minerals from the plants and animals they consume
5. The *sulfur cycle* comprises both sedimentary and gaseous portions
    a. The sedimentary portion of the cycle consists of long-term storage of sulfur within the soil and sediments
    b. The gaseous phase of the cycle consists of hydrogen sulfide ($H_2S$) and sulfur dioxide ($SO_2$) gases within the atmosphere
    c. $H_2S$ enters the atmosphere from combustion of fossil fuels, volcanic activity, and decomposition in both aquatic and terrestrial ecosystems
    d. Once in the atmosphere, $H_2S$ rapidly oxidizes to $SO_2$ and sulfate ($SO_4$)
    e. $SO_2$ and $SO_4$ are soluble in water and fall to the earth's surface in precipitation
    f. $SO_2$ and $SO_4$ are taken up by plants and incorporated into sulfur-bearing amino acids in proteins
    g. Once incorporated into plant proteins, sulfur is transferred through the food chain from consumer to consumer
    h. Sulfur in living organisms is returned to the soil by excretion and decay, where microorganisms release it as hydrogen sulfur or sulfate
    i. Bacteria play a major role in the sulfur cycle
      (1) Purple sulfur bacteria produce sulfate ions during photosynthesis without oxygen

(2) Desulfovibrio bacteria release $H_2S$ under anaerobic conditions; they are found in the mud of bogs, lakes, and estuaries where oxygen levels are very low
6. The *phosphorus cycle* differs from the sulfur cycle in that phosphorus, unlike sulfur, is not present in any form in the atmosphere
   a. Under natural conditions, phosphorus typically is in short supply
      (1) Phosphorus is soluble only under acidic conditions
      (2) In soil, phosphorus readily becomes immobilized as it forms compounds with calcium or iron
   b. The major reservoir compartment for phosphorus is rock and natural geological formations or deposits, from which the element is released by weathering, leaching, and erosion
   c. Phosphorus passes through terrestrial and aquatic ecosystems by way of plants, herbivores, and other consumers; it returns to the soil through excretion and death and decay
   d. Seabirds apparently have played an important role in the cycle, moving phosphorus from sea to land as they consume marine fish and deposit excrement (guano) on land areas where they roost
   e. Phosphorus is a major factor in the enrichment of aquatic systems in which explosive growth of algae directly results from heavy discharges of phosphorus from terrestrial ecosystems or from anthropogenic (that is, processed or generated by human activity) sources, such as sewage treatment plants
   f. Part of the phosphorus in aquatic ecosystems is deposited in sediments; some of this is effectively removed from the cycle for extended periods until a major geological event exposes these sediments to weathering

## II. Human Intrusions and Disruptions

### A. General information
1. Human intrusion into the biogeochemical cycles often results in disruptions of the normal flow of materials within an ecosystem
   a. The disruptions may occur because human activities "overload" certain portions of a natural cycle
   b. Disruptions to the biogeochemical cycles also may occur when human activities "redirect" material flow through an ecosystem and therefore disrupt the natural course of events
2. Human activities, such as mining, burning of fossil fuels, fertilizing of agricultural crops, and releasing toxins into the environment, often result in changes in the natural movement of materials within the trophic structure of ecosystems and can result in damage to both natural and human systems

### B. Heavy metals, pesticides, and other toxins
1. Human activities have increased the concentration of certain heavy metals and toxic chemicals in the environment; in turn, these substances pass through the trophic levels of ecosystems
2. In a process known as ***biological magnification*** (or biological amplification), concentrations of certain chemicals in organisms at high trophic levels in a

food chain are significantly higher than concentrations of those chemicals in organisms at lower trophic levels
   a. Chemicals subject to biological magnification generally are *nonbiodegradable* or slowly biodegradable (capable of natural decomposition) and therefore persist for a long time in the ecosystem
   b. Compounds that accumulate in higher trophic levels also are generally soluble in the body fat (lipid compounds) of organisms, which results in the substance remaining in the organism once it enters the body; for example, the fat-soluble substances are less likely to be "washed" from the body in urine and fecal material and consequently tend to build up in the organism (examples include many hydrocarbon-based pesticides and some metals)
   c. Many of the substances that are biologically magnified also adhere readily to **detritus** and soil particles, a property that increases the duration of exposure
   d. The accumulation of materials in a food chain is directly related to the concept of ecological efficiency; because much energy is lost in the transfer from one trophic level to the next, organisms of one trophic level must consume and process a large amount of the trophic level immediately below it in the food chain and, therefore, they consume large quantities of any toxic contaminant
3. A classic example of biological magnification is provided by the pesticide dichlorodiphenyltrichloroethane (DDT), which was widely used in the United States before it was banned in 1972 (it no longer is used in the United States, Canada, and Europe but still is used in other parts of the world)
   a. Evidence of biological magnification of DDT in aquatic food chains began to accumulate in the 1950s and 1960s, when populations of fish-eating birds (for example, osprey, cormorant, pelican, and bald eagle) started to decline
   b. These fish-eating birds are near the top of the aquatic food chains, and thus they ingested large amounts of the biologically magnified DDT
   c. DDT is fat-soluble, can persist for nearly 20 years, and is chemically degraded to another toxic chemical, DDE, which reduces the amount of calcium in the shells of bird eggs, causing the shells to be so thin that they break and the chicks die
   d. The biological magnification of DDT was well documented in the estuary food chains of Long Island Sound (see *DDT Concentration in the Long Island Sound Food Chain*)
      (1) DDT was sprayed in marshy areas adjacent to Long Island Sound to control insect populations
      (2) Tracing DDT through the food chain demonstrated an increase in concentration at every trophic level
   e. Accumulation of high concentrations of DDT in tissues can result in death, impaired reproduction, or disruption of the genetic constitution of organisms
      (1) Laboratory investigations have demonstrated that zooplankton, shrimp, and crabs are killed outright by exposure to DDT in concentrations of only a few parts per billion
      (2) Freshwater trout with ovarian concentrations of 5.0 parts per million (ppm) of DDT suffer nearly 100% mortality of their young fry

## DDT Concentration in the Long Island Sound Food Chain

Prior to the U.S. ban on the use of DDT, it was used as an effective pesticide to control insect populations in many areas. DDT was found to accumulate in increasing concentration within many food chains. The food chain of Long Island Sound was extensively studied to determine the pathway and concentrations of DDT at each trophic level.

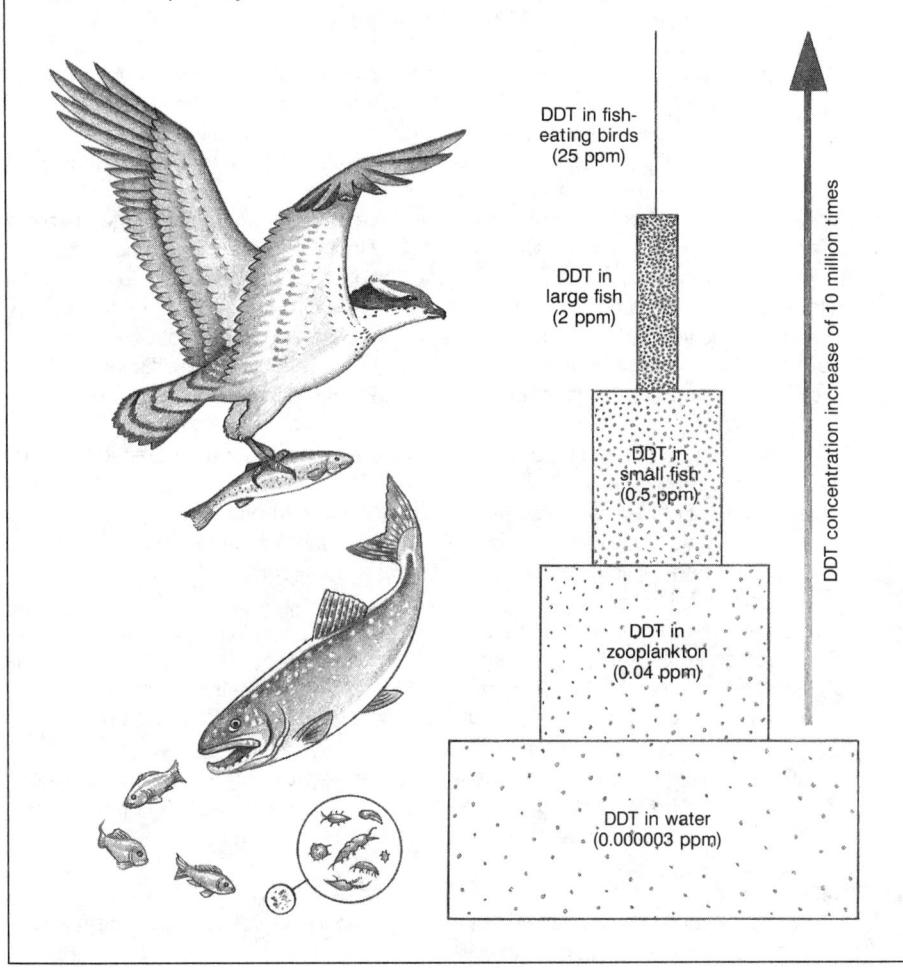

DDT in fish-eating birds (25 ppm)
DDT in large fish (2 ppm)
DDT in small fish (0.5 ppm)
DDT in zooplankton (0.04 ppm)
DDT in water (0.000003 ppm)

DDT concentration increase of 10 million times

4. Other substances that are biologically magnified in food chains include heavy metals, such as lead and mercury, and polychlorinated biphenyls (PCBs) used in numerous industrial processes

### C. Ozone shield depletion
1. The *ozone shield* or layer is a layer of ozone (a molecular form of oxygen $O_3$ having marked oxidizing capability) located at altitudes of 6 to 10 miles up in the atmosphere
2. The ozone shield is ecologically important because it removes significant amounts of potentially damaging ultraviolet radiation from sunlight, thus protecting the earth's biosphere from overexposure to ultraviolet radiation

a. Loss of this protective shield can have human health implications
      (1) A 1% loss in the ozone layer can result in a 2% increase in ultraviolet light reaching the earth's surface, which can result in a 5% to 7% increase in the incidence of skin cancer
      (2) Exposure to higher levels of ultraviolet light also has been implicated in increased incidence of cataracts and a general weakening of the body's immune system
   b. Increased exposure to ultraviolet light also may damage photosynthetic plankton species in oceans and thereby disrupt marine food chains
   c. Excessive ultraviolet light also damages the photosynthetic machinery of terrestrial plants, with resulting negative impacts on human food production and terrestrial food chains
3. A number of chemical reactions that take place in the atmosphere consume ozone and reform oxygen; these reactions typically involve gases containing nitrogen and hydrogen, both of which are naturally plentiful in the atmosphere
4. Under natural conditions there is a balance between the rate of ozone formation and destruction that results in a more or less constant amount of ozone
5. In recent history, human activity has released a number of chemicals into the global atmosphere that destroy ozone and upset the natural equilibrium of ozone in the upper atmosphere
   a. The major culprit in ozone depletion is the element chlorine released into the upper atmosphere
   b. Sources of chlorine release include chlorofluorocarbons (in refrigerants, industrial solvents, spray propellants), carbon tetrachloride (a cleaning agent and industrial solvent), and nitric oxides (combustion waste products from fossil fuels)
   c. Once these products are released and decompose within the atmosphere to release chlorine, the attack on the ozone begins
   d. A chlorine atom acts as a catalyst and is capable of destroying many molecules of ozone before it chemically reacts with other substances or is otherwise removed from the atmosphere
   e. Global levels of ozone have not yet dropped appreciably, but levels over the Antarctic have dropped by nearly 40%, and some investigators indicate a 2.3% to 6.0% decline over the United States

**D. Global warming**
1. On a global scale, carbon dioxide in the atmosphere serves to hold in heat and keep it from escaping into space
2. As first pointed out by G.E. Hutchinson in 1949, there is the distinct possibility of perturbation of atmospheric carbon dioxide levels caused by industrial activity or, more likely, caused by the destruction of forests and bogs, resulting in increased release of carbon dioxide
3. Subsequent data suggest that there has been a sustained and significant increase in the concentration of atmospheric carbon dioxide
   a. Many scientists believe that both industrial output and deforestation are responsible for the rise in carbon dioxide
   b. Some investigators contend that about half of the increment in carbon dioxide floating in the atmosphere comes from fossil fuel combustion
   c. Other gases thought to contribute to global warming include nitrous oxides, methane, ozone, and chlorofluorocarbons

4. In the short term, the level of carbon dioxide can vary — particularly on an annual or seasonal basis, as influenced by photosynthesis in temperate climates
    a. In the summer, vegetation takes in more carbon dioxide than it releases, and so the atmospheric carbon dioxide level declines
    b. During the winter, plant respiration releases more carbon dioxide than photosynthesis can assimilate; consequently, the atmospheric carbon dioxide level increases
5. The oceans may provide some buffering of atmospheric carbon dioxide levels but are likely to be ineffective at buffering large changes, because carbon dioxide readily diffuses into and out of upper ocean layers but not ocean depths
    a. The concentration of carbon dioxide in the surface waters is influenced by the concentration of carbon dioxide in the air, by temperature and ocean chemistry, and by biological processes
    b. Water temperature is extremely important in determining carbon dioxide retention; solubility of carbon dioxide falls as temperature increases
    c. The ocean's role in atmospheric carbon dioxide balance is not fully understood; it is quite possible that as oceans warm they will be sources of additional carbon dioxide
6. The burning of fossil fuels has been implicated as a source of increased atmospheric carbon dioxide
    a. The amount of carbon dioxide released from fossil fuels so far is equal to 28% of the current atmospheric reserve
    b. It is estimated that human activities could potentially release about 3 or 4 atmospheres' worth of carbon dioxide — thus far, most authorities estimate that human activities have released only one-quarter of one atmosphere
7. Clearing forests, burning peat, draining marshes, and plowing prairies all release carbon dioxide into the atmosphere; it is believed that humans are removing these areas faster than they can be replaced, and thus there may be a significant increase in the net release of carbon dioxide
8. In such a complex situation as global warming, various complicating factors may alter the overall trends
    a. When a forest is cleared, ecological succession is reset to an earlier stage; the plants in these earlier stages tend to have rapid growth and thus maximize the influx of carbon dioxide into plants and remove it from the air
    b. The rate of photosynthesis can be altered by the concentration of carbon dioxide, and the release of carbon dioxide into the atmosphere may enhance photosynthesis
    c. Human activities and natural events (such as volcanic eruptions and dust storms) can add particulate material to the atmosphere, which can, in turn, reflect more of the sun's warming rays back into space and thus work to counteract the global warming phenomenon
9. The major concern related to global warming is the possible disruption of global wind and the climatic patterns that may result — the emphasis is on potential climate changes rather than merely a warming trend
    a. The warming patterns on a global scale will not be uniform, but rather there will be a decline in the overall temperature difference between the poles and the equator, which might result in dramatic changes of the prevailing wind patterns and consequently alter precipitation patterns
    b. Warming of polar regions also might melt polar ice caps, causing a rise in sea level and subsequent coastal flooding

## E. Acid precipitation

1. As a result of the formation of carbonic acid from carbon dioxide and water present in the atmosphere, uncontaminated precipitation typically has a pH of about 5.6; *acid precipitation* describes precipitation with a pH of less than 5.6
   a. The amounts of sulfur dioxide and nitrogen oxides injected into the atmosphere have increased significantly in the last century
   b. Once in the atmosphere, these materials become involved in complex chemical reactions that transform them into nitric acid, sulfuric acid, and sulfate and nitrate salts, which alter the pH of precipitation
2. The major human sources of sulfur and nitrogen compounds that contribute to acid precipitation are internal combustion engines, agricultural fertilizers, and coal-fired electric generating plants
3. During precipitation events, these acids fall to the earth's surface, where they affect aquatic, terrestrial, and human ecosystems
   a. Lakes and streams differ in their ability to absorb and buffer the input of acid precipitation; however, as pH approaches 4.0, aquatic organisms become stressed and some are eliminated
      (1) The effects of acid precipitation on aquatic systems seldom result in direct killing of adults but often interfere with successful reproduction
      (2) The acidification of aquatic ecosystems also often decreases algae and zooplankton growth, both of which are near the base of the food chain; this disruption can be very damaging to the ecosystem
      (3) Acidic waters also leach aluminum, lead, and mercury from environmental surfaces (for example, mine spoils, roofs, contaminated soils); these metals can be toxic to aquatic life
   b. The effects of acid precipitation on forest ecosystems are more difficult to substantiate than are the effects on aquatic ecosystems
      (1) Acidic precipitation has been shown to damage the outer protective layers of plant organs (the cuticle of leaves and stems), making them more susceptible to disease and insects
      (2) Leaching of mineral nutrients from soils and from leaf tissue also has been shown to occur when precipitation becomes acidic
      (3) Plants also will take up toxic substances from soils when the toxins are mobilized by a lowered pH
      (4) Plants are the primary producers of forest ecosystems and, as such, any negative impact on them may be detrimental to the forest ecosystem
   c. Acid precipitation potentially could affect humans in several ways
      (1) Acid precipitation can deteriorate or corrode metals, paints, and concrete, with a resultant economic impact
      (2) Acid precipitation can affect human health indirectly by potentially damaging food crops (diminishing plant growth or making crops more susceptible to pests and disease) and thus affecting the food supply
      (3) Acidification of drinking water supplies can pose a potential hazard as toxins are mobilized and transported in the drinking water

## Study Activities

1. Choose a sedimentary and a gaseous biogeochemical cycle and sketch the major flows through each cycle.
2. Define nitrogen fixation, ammonification, denitrification, and nitrification; describe their importance in terms of the global cycling of nitrogen.
3. Sketch a food chain and explain how a toxic substance may accumulate in various trophic levels.
4. Choose one of the global environmental pollution problems discussed in this chapter, and write a short essay describing its causes and possible ecological effects.

# 6

# Population Ecology Basics

## Objectives

After studying this chapter, the reader should be able to:
- Explain the difference between crude density and ecological density.
- Describe the various patterns of dispersion of organisms and the conditions that lead to such dispersions.
- Describe the concept of natality and its role in population growth.
- Explain the importance of mortality, mortality rate, birth rate, life expectancy, and survivorship.
- Compare the three theoretical forms of survivorship curves.
- Explain the implications of age structure for future trends in population growth.
- Describe the concept of biotic potential.
- Compare and contrast exponential and logistic models of population growth.
- Explain the significance of a life table.
- Describe the impact of density-dependent and density-independent factors on population growth.

## I. Properties of Populations

### A. General information
1. Populations are groups of individuals of the same species living in the same location at the same time
2. Populations often have traits different from those of the individuals that comprise them; an individual is born and dies once, but a population continues, changing in size based on birth and death rates
3. Populations can be described and characterized by a variety of measures
   a. Density and dispersion help ecologists to characterize the distribution of a population
   b. Natality, mortality, survival, and life tables enable ecologists to describe the birth, death, and length of life of individuals within a population
   c. Age structure and sex ratios help ecologists characterize the members of the population with respect to age and sex

### B. Density and dispersion
1. *Density* refers to the number of individuals per unit area

a. Density (more specifically, *crude density*) generally is expressed as the number of individuals, or population biomass, per unit area or volume; for example, 200 trees per acre or 200 pounds of fish per acre of surface water
   b. Ecologists sometimes limit their measure of density to the habitat space actually inhabited or utilized by the species; this concept is termed **ecological density**
 2. *Dispersion* is the distribution of individuals of a population over a given geographical area or distribution of individuals of a population over time
   a. Organisms may be distributed throughout the physical environment in a random, uniform, or clumped manner
      (1) *Random distribution* occurs when the position or location of one organism is independent of others; it generally occurs only in uniform habitats and so is very rare, but is seen in some forest invertebrates, such as spiders
      (2) *Uniform,* or *regular, distribution* produces even spacing of individuals; it often occurs as a result of intraspecific competition events, such as establishment of a territory
      (3) *Clumped,* or *cluster, distribution,* the most common type of distribution, is generated by habitat differences, such as the distribution of food supply or seasonal weather conditions
      (4) The dispersion pattern of a population is likely to be some variant of these three types and is unlikely to be purely random, uniform, or clumped
   b. *Temporal dispersion* refers to the distribution of organisms of a population over time; it can be considered **circadian** (relating to daily changes in night and day) or *seasonal* (based on the organism's life cycle)

## C. Natality
 1. New individuals can be added to a population by a reproductive event or by immigration
 2. Not all organisms are born (some are hatched, germinated, or produced by nonsexual means); the production of new individuals in a population is termed **natality**
 3. Natality in animals generally is measured as crude birth rate or specific birth rate
    a. *Crude birth rate* is the number of births expressed in terms of population size, such as number of offspring per 1,000 individuals
    b. *Specific birth rate* is the number of births expressed in relation to age; it is equal to the number of offspring produced per unit of time by females in different age classes
    c. Because the rate of population increase in sexually reproducing animals depends on the number of females in the population, age-specific birth rate counts only females giving rise to females; males are not considered
 4. Natality in plants is more difficult to define and measure because plants produce new structural units (that is, new buds, leaves, flowers, fruits, and roots) that make it difficult to identify and count individuals in the population
    a. Each structural unit of a plant has its own natality rate
    b. Most plants produce seeds as a result of sexual reproduction; seed production further complicates the examination of plant natality
       (1) With the exception of annual and biennial plants, it is difficult to estimate seed production by individual plants and by populations because per-

ennial plants vary in longevity and seed production, and seed production can vary significantly from year to year
- (2) Many seeds undergo a period of dormancy, which may last for several years in some species
- (3) As a result of seed dormancy and variable seed production, seeds may accumulate in the soil to germinate at some later date; this "storehouse" of seeds in the soil, known as a **seed bank,** may play a significant role in the life cycles of many plant species
- (4) As a general rule, seeds that have periods of dormancy and make a significant contribution to seed banks more commonly come from annuals and short-lived species than from long-lived species
- (5) Plants that might otherwise be classified as annuals but have significant seed banks cannot accurately be described as true annuals, because the "effective life" of the annual plant is extended beyond a year by seeds that can remain viable for longer than a year and germinate when conditions are favorable

## D. Mortality and survivorship
1. **Mortality,** the death of individuals in a population, can be considered the opposite of natality
2. Mortality is expressed as either mortality rate or death rate
    a. **Mortality rate,** defined as the probability of dying, often is measured as the number of individuals that died during a given time interval divided by the number that were alive at the beginning of the interval
    b. **Death rate** is defined as the number of deaths that occur during a given time interval divided by the average population size during that interval
3. The counterpart of mortality rate is the probability of living, or *survivorship,* which is defined as the number of survivors divided by the number alive at the beginning of a specific time period
4. Because the number of individuals alive in a population is more important than the number dying, ecologists often discuss survivorship rather than mortality
5. Another measure of survivorship often considered by ecologists is **life expectancy,** the average number of remaining years to be lived by members of a population

## E. Life table
1. A *life table* is a tabular representation of mortality, survivorship, and life expectancy for a population (see *Human Life Table*)
    a. A typical life table consists of a series of columns, each of which describes a particular aspect of the population according to age
    b. For clarity, the data are organized into groups of specific age classes known as **cohorts**
2. Mortality and survivorship often are expressed graphically to visualize the overall trends in a population (see *Mortality and Survivorship Curves,* page 56)
    a. A *mortality curve* is derived by plotting the mortality rate of a cohort ($q_x$ from a life table) against the age of all members of the population (starting at zero — birth — and ending with the maximum age)
        (1) A mortality curve generally consists of two parts: juvenile phase and postjuvenile phase

## Human Life Table

The numbers within the life table are for a cohort or group of individuals of the same age or age class. Data are derived from a partial survey of Sleepy Hollow Cemetery in Tarrytown, New York.

| $x$ | $l_x$ | $d_x$ | $q_x$ | $e_x$ |
|---|---|---|---|---|
| 0 | 1,222 | 31 | .025 | 59.9 |
| 1 | 1,191 | 39 | .033 | 60.4 |
| 6 | 1,152 | 15 | .013 | 57.4 |
| 11 | 1,137 | 13 | .011 | 53.1 |
| 16 | 1,124 | 32 | .028 | 48.7 |
| 21 | 1,092 | 22 | .020 | 45.1 |
| 26 | 1,070 | 36 | .034 | 40.9 |
| 31 | 1,034 | 37 | .036 | 37.3 |
| 36 | 997 | 45 | .045 | 33.6 |
| 41 | 952 | 35 | .037 | 30.0 |
| 46 | 917 | 48 | .052 | 26.7 |
| 51 | 869 | 53 | .061 | 23.5 |
| 56 | 816 | 77 | .094 | 19.9 |
| 61 | 739 | 93 | .126 | 16.7 |
| 66 | 646 | 117 | .181 | 13.7 |
| 71 | 529 | 120 | .227 | 11.2 |
| 76 | 409 | 137 | .335 | 8.7 |
| 81 | 272 | 127 | .467 | 6.9 |
| 86 | 145 | 79 | .545 | 5.7 |
| 91 | 66 | 45 | .682 | 4.6 |
| 96 | 21 | 16 | .762 | 4.0 |
| 101 | 5 | 5 | 1.000 | 4.0 |

**KEY**
$x$ = age
$l_x$ = number of individuals in a cohort that survive the listed age level
$d_x$ = number of individuals in a cohort that die in an age interval $x$ to $x + 1$, from one age level listed to the next
$q_x$ = probability of dying or age-specific mortality rate, determined by dividing the number of individuals that died during the age interval by the number of individuals alive at the beginning of the age interval
$e_x$ = life expectancy (number of years left to live)

  (2) Mortality during the juvenile phase typically is high; mortality during the postjuvenile phase initially declines with age and then increases again as the individuals reach the maximum length of their life span
 b. A *survivorship curve* is derived by plotting survivorship of a cohort ($l_x$ from a life table) against age of all members in the population (starting at zero and ending with the maximum age)

## Mortality and Survivorship Curves

A mortality curve (see illustration A) is obtained by plotting mortality $(q_x)$ against age $(x)$ from a life table. The characteristic curve for most organisms is shown here and is referred to as a J-shaped curve. A survivorship curve (see illustration B) is obtained by plotting the survivorship $(l_x)$ against the age $(x)$ from a life table. There are three hypothetical shapes to the survivorship curves, designated types I, II, and III.

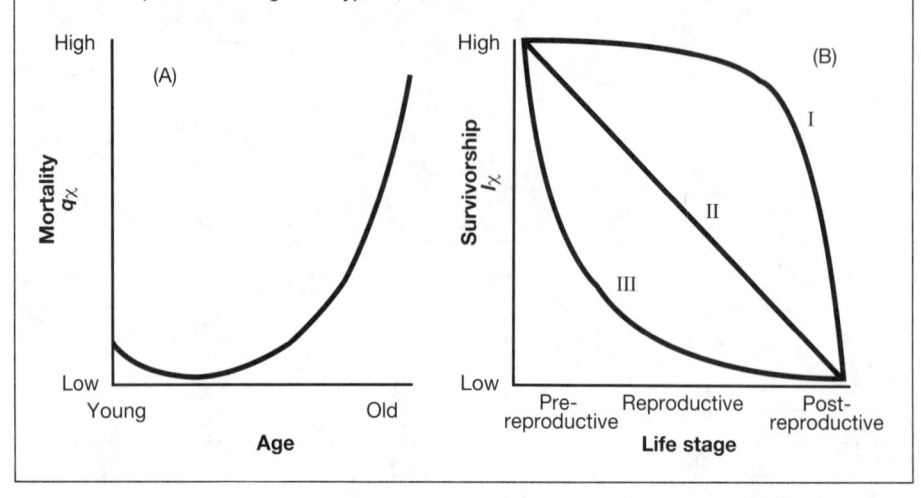

(1) Survivorship curves are classified into three hypothetical classes, designated type I, II, and III, and are characterized as possessing typical shapes
   (a) A type I survivorship curve has a convex shape and is typical of organisms that live out most of their physiological life span (that is, die of old age), such as many mammals, some plants, and humans
   (b) A type II survivorship curve has a linear appearance and is typical of organisms with more or less constant mortality rates, such as adult birds, rodents, and some plants
   (c) A type III survivorship curve has a concave shape and is typical of organisms with high early mortality, such as invertebrates, fish, and some plants
(2) The three types of survivorship curves are conceptual models and not classes to which any particular survivorship must conform

## F. Age structure
1. **Age structure** refers to the classification of a population in cohorts (age groups) to examine its life history
2. Typically, the members of a population are categorized into three classes: prereproductive, reproductive, and postreproductive
3. If the proportion of individuals of a population in each category is portrayed graphically, one can construct an age structure diagram
4. Age structure diagrams are constructed so that the prereproductive class of the population forms the base of the diagram, the reproductive class forms the midsection, and the postreproductive class forms the top

a. The shape of an age structure diagram can be a predictor of future population trends
   (1) If the age structure diagram is triangular, there is potential future growth in the population because of the disproportionate number of juveniles that will someday become reproducing adults
   (2) If the age structure diagram is rectangular or bullet-shaped, the population is considered to be of stable size and there is likely to be little or no growth
   (3) If the age structure diagram forms an inverted triangle, the population is likely to decline
b. Age structure diagrams are merely "snapshots" in time of a population; a particular population may not exactly fit one of the three shapes described above

## G. Sex ratio
1. The *sex ratio* of a population is the ratio of males to females
2. Populations of many organisms tend to have sex ratios of 1:1, in which the proportion of males and females is equal
3. Sex ratios often are considered to be primary or secondary
   a. The *primary sex ratio* is the ratio of males to females at conception
   b. The *secondary sex ratio* is the ratio at birth
   c. The difference between primary and secondary sex ratios is indicative of differences in mortality between males and females during *gestation,* the period during which the offspring is carried within the maternal body
4. In many mammals, the primary and initial secondary sex ratios tend to be weighted toward males
   a. For example, in humans the primary sex ratio is 114:100, the secondary sex ratio is 106:100, the ratio at age 40 to 44 is 100:100, the ratio at age 60 is 88:100, and the ratio at age 80 to 84 is 54:100
   b. This trend in sex ratios is the result of lower mortality in females

## II. Population Growth and Regulation

### A. General information
1. Populations are dynamic and may change over a variety of time frames (that is, hourly, daily, seasonally, annually)
2. Local populations may expand, contract, or disappear while the species as a whole remains unchanged
3. Changes in populations occur through changes in birth rates, death rates, and movement of individuals in and out of a particular population
4. Because population characteristics (natality, mortality, survivorship) often change continuously, ecologists are interested in rates of population change rather than simply indicators of change
5. The study of population characteristics and changes is referred to as *demography*

### B. Population increase
1. The rate of population increase, or **biotic potential,** is a measure of a population's ability to increase in size

a. Biotic potential, often represented by $r$, refers to the intrinsic rate of natural increase of a population
b. The intrinsic rate of natural increase is an instantaneous measure of population increase, determined by both birth rates and death rates
c. Mathematically, biotic potential can be represented by the equation:
   $r = b - d$, where b equals birth rate and d equals death rate
d. Biotic potential is the measure of increase per individual member of the population when the population is in an uncrowded environment
e. As the term suggests, much of the biotic potential is growth that can be realized only under certain (uncrowded) conditions
f. If two populations with different biotic potentials experience some catastrophic decline, the population with the higher biotic potential will be able to rebound more quickly

2. Certain characteristics of a population tend to result in a higher biotic potential
   a. Populations with high biotic potential generally produce more offspring per breeding period than do populations with lower biotic potential; they tend to produce larger litters, more seeds per plant, more eggs per clutch, and so forth
   b. The survival of offspring up to and through the reproductive age span tends to be characteristic of populations with high biotic potential
   c. Populations with high biotic potential tend to have longer reproductive age spans
      (1) **Semelparous** species have only one reproductive period during their lifetime and typically die after reproducing; examples include salmon and annual plants that produce seeds at the end of their growing season
      (2) **Iteroparous** species have several reproductive periods during their lifetime or have long reproductive periods during which they can repeatedly produce offspring; examples include humans, most other mammals, and oak trees, which may produce acorns numerous times during their life
   d. Populations with high biotic potential usually reproduce at a younger age
      (1) The age of first reproduction is considered the most important characteristic influencing biotic potential
      (2) A species with an earlier age of first reproduction will be able to produce offspring sooner, and those offspring will reach reproductive age sooner; the overall result is a greater potential for population increase

3. The exact determination of biotic potential is tedious and time-consuming, but some measurements have been made
   a. Large mammals, such as elephants, rhinoceroses, and humans, typically have a biotic potential between 0.02 and 0.5 per year
   b. The biotic potentials for bird species range from 0.05 to 1.5 per year
   c. Small mammals, such as squirrels and rabbits, have biotic potentials of 0.3 to 8.0 per year
   d. Insects have biotic potentials of 4.0 to 50.0 per year
   e. Bacteria have biotic potentials of 3,000 to 20,000 per year

## C. Models of population growth
1. To examine the growth of populations, scientists generally recognize two models of population: exponential and logistic (see *Exponential and Logistic Growth Curves*, page 60)
2. Discussions of population growth models can become very mathematical; here an attempt will be made to graphically analyze the major aspects of the two models, and the mathematical equations will be de-emphasized
3. The *exponential growth model* is characteristic of populations presented with an unlimited environment and a small initial population size
    a. Exponential growth may occur for some time when a small number of organisms enters a suitable but previously unoccupied habitat
    b. For exponential growth to be sustained, there must be no limitations on growth — in other words, unlimited space, no overcrowding, and limitless resources
    c. The number of reproducing individuals steadily increases, and the growth rate steadily increases from an initial slow rate (when population size is small) to a rapid rate (when population size is large); graphically, this produces a J-shaped curve when the number of individuals is plotted against time
    d. Exponential growth generally is obtained only in experimental or laboratory conditions; it occurs in nature only occasionally
    e. The mathematical equation representing exponential growth is written as: $dN/dt = rN$, where $dN$ represents the change in population size, $dt$ represents the time interval appropriate to the species' life span and generation time over which the population growth is being considered, $r$ represents rate of population growth, and $N$ represents the population size
4. The *logistic growth model* takes into account the limits on growth experienced by populations in nature
    a. In general, the overall shape of the logistic curve is sigmoid, or S-shaped
    b. Logistic growth is characterized by an initial exponential phase, followed by a slowing growth rate; eventually, the growth rate reaches zero and balances with the death rate
    c. Factors that tend to slow population growth (for example, the effects of crowding, which result in lowered reproduction, poor nutrition, high death rate caused by predation, increased emigration, and competition) are termed **environmental resistance**
    d. Environmental resistance increases as the population size increases, putting the "brakes" on population increase
        (1) When population size is small, environmental resistance is slight; thus biotic potential is realized and growth is rapid
        (2) When population size is large, environmental resistance becomes a determining factor; the growth rate declines and may even become negative (a negative growth rate means the population is declining)
    e. The logistic model of population growth introduces the concept of **carrying capacity** (often designated as $K$), the maximum size of a population that an area can support
    f. Some populations demonstrate growth curves that appear similar to those described by the logistic equation but that overshoot $K$ and therefore exceed the carrying capacity

## Exponential and Logistic Growth Curves

Exponential and logistic forms of population growth are presented graphically. Exponential growth occurs when a population experiences no limits to growth, as shown in the diagram on the left. In logistic growth, the population starts out very small and grows slowly; the growth accelerates, then decelerates as the population reaches the maximum size or carrying capacity (this is shown in the diagram on the right).

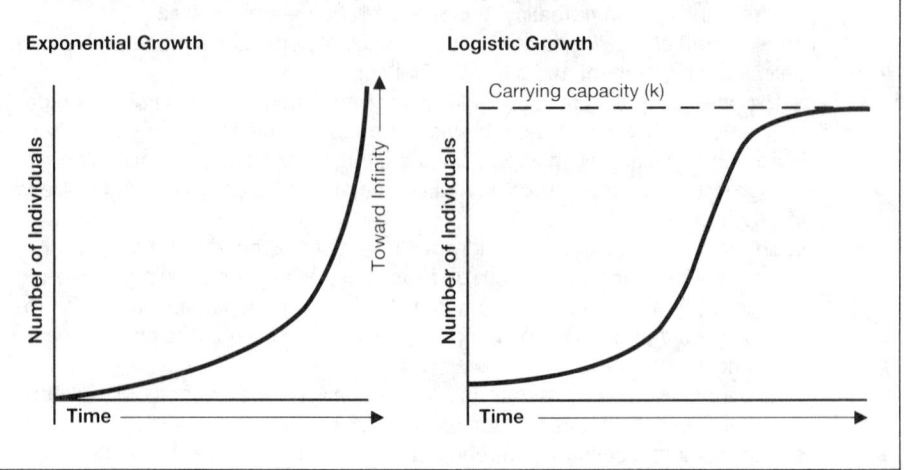

**Exponential Growth**

(Number of Individuals vs. Time — curve rising toward infinity)

**Logistic Growth**

(Number of Individuals vs. Time — S-shaped curve approaching Carrying capacity (k))

   (1) Overshooting carrying capacity may be caused by a delay between the time a population reaches a certain size and the time at which unfavorable conditions caused by overcrowding are felt by reproductive members of the population
   (2) If a population exceeds the carrying capacity of its environment, then the population will decrease; this decrease often is the result of increased death rates
   (3) A population that has exceeded the carrying capacity also is in danger of extinction because the population may severely stress the environment, so that a total collapse is inevitable
g. Although the logistic model better describes the natural growth of populations, it does not account for some limitations and conditions
   (1) The environmental resistance concept of logistic growth does not account for the positive aspects of larger population size, such as enhanced detection of food, predators, or prey or the ability of groups of organisms to collectively modify their immediate environment in a favorable way (for example, when animals huddle together to keep warm)
   (2) The logistic model does not mathematically account for time lags or delays, which often are inherent in natural populations
   (3) The logistic model treats all individuals of a population as equivalent; however, there are likely to be differences in age, reproductive status, or physical size — all of which could impact population growth
h. The mathematical equation representing logistic growth is written as: $dN/dt = rN[(K - N)/K]$, where dN represents the change in population size, dt represents the time interval appropriate to the species' life span and generation time over which the population growth is being considered, $r$

represents rate of population growth, N represents the population size, and K represents carrying capacity for the species
  (1) The logistic equation is constructed by starting with the model of exponential population growth and creating a term that reduces the value of r as the population size increases
  (2) If K is the carrying capacity, the term $(K - N)$ indicates how many new individuals the environment can accommodate
  (3) The term $(K - N)/K$ indicates what percentage of K still is available for population growth
  (4) Note that when N is low, the term $(K - N)/K$ is large and r is not diminished; but when N is large and resources become limiting, the term $(K - N)/K$ will be small and r is substantially reduced and thus the growth of the population slows

## D. Extinction
1. Extinction can occur when the biotic potential r becomes negative and the population cannot recover
2. Some populations are more susceptible to extinction than others
    a. Organisms with a large body size are likely to be in more danger of extinction because they require more food and space in which to survive
    b. Organisms with a small, restricted, or specialized geographic range also are more susceptible to extinction because their habitat could easily be destroyed
    c. Populations in which genetic variety has been reduced or restricted are less able to adapt to change and consequently are more likely to become extinct
3. Once a population has declined significantly, it may be difficult or impossible for it to rebound
    a. In sparse populations, the females of reproductive age may have only a small chance of meeting males for successful reproduction, further reducing the potential for growth
    b. Sparse populations may not have sufficient numbers to stimulate the social behavior necessary for successful reproduction
4. Major causes of extinction generally are related to human activities
    a. More than 33% of the mammals and 43% of the birds known to have become extinct since the year 1600 were eliminated by hunting
    b. With continued human pressures, the more recent trend in the causes of extinction is habitat destruction

## E. Density-dependent regulation
1. *Density-dependent* factors affect a population in proportion to its size; at low density, mortality is independent of population size, and at higher density, mortality increases and population growth is slowed
2. Density-dependent factors impact mortality and natality through resource shortages and competition among members of the population to obtain those resources
3. Examples of density-dependent factors include predation, parasitism, disease, and competition

## F. Density-independent regulation
1. *Density-independent* factors affect the same proportion of individuals in a population regardless of the population density
2. Density-independent factors often are related to climate; examples include rapid temperature changes, drought conditions, and flooding
3. In most cases, density-independent factors by themselves do not really regulate population size; rather, they simply alter the number of individuals by affecting birth rate or death rate

## G. Fluctuations and cycles
1. Some populations oscillate between high and low points at a more or less regular interval; the most common intervals between peaks of population size are 3 to 4 years and 9 to 10 years
   a. Lemmings demonstrate a typical cycle of population peaks every 3 to 4 years, when mass migration occurs
   b. Snowshoe hare populations display a typical population cycle of 9 to 10 years between peaks
2. These fluctuations often occur as a result of density-dependent mechanisms, which in turn alter birth and death rates, resulting in a regular oscillation of population size

---

# Study Activities

1. Sketch and label random, uniform, and clumped dispersion or density patterns.
2. Sketch the general form of the three theoretical survivorship curves.
3. Draw a hypothetical age structure diagram of a population that is stable and not likely to show great change in population size.
4. Draw graphs of exponential and logistic population growth; then explain what the graphs represent.
5. List the characteristics of a species that put it at high risk for extinction.
6. List the characteristics of a species likely to result in a high biotic potential.

# 7

# Competition and Other Population Interactions

## Objectives

After studying this chapter, the reader should be able to:
- Name, explain, and give examples of the various ways in which populations interact with one another.
- Differentiate between scramble and contest competition.
- Describe the role of density and stress in controlling the size of a population.
- Discuss the role of dispersal in a population.
- Explain the role of social dominance in population biology.
- Define territoriality and describe its importance as a social interaction.
- Define niche theory and explain its implications with respect to interspecific competition.

## I. Population Interactions

### A. General information
1. Individuals of a species interact with individuals of their own species and with individuals of other species
2. Within a community there is a range of interactions among the various groups; some populations have minimal influence on one another, whereas others have direct and immediate impacts on one another

### B. Types of population interactions
1. Interactions among populations can have positive, negative, or neutral effects on the growth of populations involved
2. **Symbiosis** is an interaction between two species in direct contact with one another; it encompasses three types of interactions: mutualism, commensalism, and parasitism
   a. **Mutualism** is an interaction between populations in which both populations benefit; examples include photosynthesis of unicellular algae in tissues of corals, association of fungi with the roots of some plants, and interactions of pollinators with specific species of flowering plants
   b. **Commensalism** is a one-sided relationship between two species in which one benefits and the other neither benefits nor is harmed; examples include cattle egrets (heronlike birds) and cattle (the birds feed on insects

that cattle flush from vegetation as they graze), algae growing on the shells of aquatic turtles, and barnacles attached to whales
   c. *Parasitism* is a type of predation in which one organism consumes the other but does so in a manner such that the host (the organism being consumed) survives; examples include ticks feeding on deer and leeches feeding on aquatic mammals
3. **Amensalism** is an interaction in which one species is inhibited and the other species remains unaffected; examples include certain species of ferns and cacti that produce chemicals which inhibit the growth and survival of neighboring plants
4. **Predation** is the killing of one organism by another to obtain food; examples include coyotes preying on prairie dogs, and wolves preying on moose
5. **Competition** occurs whenever a valuable or necessary resource in short supply is sought by two or more organisms, or, if that resource is not in short supply, whenever organisms seeking the resource harm one another in the process of procuring it
   a. The competitive interaction will differ depending on the characteristics of the resource involved and of the species competing
   b. Resource or *scramble competition* (also called exploitative competition) occurs when a number of organisms deplete common resources in short supply without reducing the probability that another individual can exploit the remaining resources
      (1) Scramble competition is likely to occur if the resource is short-lived and its availability is unpredictable
      (2) The populations of species involved in scramble competition are likely to be well below the carrying capacity of the environment and probably will have a large biotic potential
   c. Interference or *contest competition* occurs when the organisms seeking a resource harm one another even if the resource is not in short supply, and access to the resource pool is reduced by pressure from or the presence of the competitor
      (1) Contest competition generally occurs when the contested resource is long-lasting and predictable and thus worth defending
      (2) The species involved in contest competition often use a chemical defense (for example, allelopathy) or a behavioral mechanism (for example, territoriality) to defend the resource
   d. Competition may be interspecific or intraspecific
      (1) *Intraspecific competition* involves interaction between members of the same species
      (2) *Interspecific competition* involves interaction between different species
      (3) Competition often increases in intensity as the similarities between the competitors increase; thus, intraspecific competition generally is considered to be more intense than interspecific competition

## II. Intraspecific Competition

### A. General information
1. Intraspecific competition affects the ability of members of a population to survive and reproduce and thus can influence population density

2. Intraspecific competition can be manifested by one organism taking as much as it can of some resource (scramble or exploitative competition), or it can involve direct interaction in which the competitor is denied access to the resource (contest or interference competition)
   3. Because intraspecific competition affects the relationship between members of the same species, its impact on the population may influence or be influenced by population density, dispersal, and social interaction

**B. Density and stress**
   1. As population density increases, the effects of stress and crowding may become apparent
      a. At higher population densities, there may be increased contact among members of a population, which may lead to stress
      b. The actual mode of action by which stress may impact a natural population is not completely understood, but many investigators believe it involves the endocrine system and thus may somehow adversely impact the physiology of the organisms
      c. Research in vertebrates indicates that increased social pressure results in stress, which triggers overactivity of the hypothalamus, pituitary gland, and adrenocortical system
         (1) Overstimulation of these hormone-regulating systems can cause dramatic changes in hormone secretion
         (2) Significant hormonal changes may result in suppressed growth, curtailed reproduction, and delayed sexual activity
         (3) Other hormonal changes may suppress the immune system and thereby make the organism more susceptible to disease
         (4) Social stress among pregnant females may increase mortality of embryos and cause inadequate milk production and subsequent stunting of offspring
   2. Increased social pressure resulting in physiological stress that affects survival and mortality in a density-dependent fashion may be important in controlling population size
   3. Plants growing in high densities also show the effects of crowding
      a. In crowded conditions, plants will respond to conditions of lowered nutrients, moisture availability, or other environmental stress
      b. Individual plants in stressful conditions respond to increased density with decreased or stunted growth
      c. Some plants respond to crowded conditions by increased seed production, with seeds dispersing to other, less crowded areas
   4. Density dependence and intraspecific competition are closely interrelated; whenever there is intraspecific competition, its effect on survival, reproduction, or a combination of the two is likely to be density-dependent

**C. Dispersal**
   1. As a result of increased density or other complex social interactions, some of the population may permanently move, or *disperse,* from an area to a new locale
   2. There is no absolute rule as to which members of a population will disperse
      a. Among bird species, the juveniles are the major class of dispersers
      b. Among many rodent species, the subadults are the predominate dispersing class

c. Among insects, dispersal often occurs among individuals of a particular phase of the life cycle — winged adults
3. Dispersal among all groups generally takes place during a prereproductive period
   a. In birds, dispersal often occurs in spring
   b. In rodents, dispersal often occurs during periods of increasing population size and high density
4. Intraspecific competition may be a driving force in the dispersal of some species
   a. Aggressive interactions among rodents during periods of high density have been implicated in forcing some members of the population to disperse
   b. No one mechanism can explain motivations for dispersal; however, dispersal often follows periods of reproduction and mutual interference, and aggressiveness often is involved
5. Currently, there is no consensus among ecologists as to whether or not dispersal can effectively regulate population size

## D. Social interaction

1. For social interactions to occur, organisms must cluster or group together into some interacting unit
   a. Groups of organisms (largely vertebrates and some insects) usually aggregate in some fashion that results in simultaneous use of space
   b. Aggregations may establish initially as a result of the distribution of a resource, in response to predation, as a consequence of patterns of dispersal, or as a result of mating systems
   c. A range or spectrum of social systems found within ecosystems is based on agonistic interactions between members of a population
      (1) *Territoriality,* one end of the spectrum or continuum of social systems, occurs when the owner or defenders of a **territory** have complete dominance and control over a particular area or portion of the habitat
      (2) At the other end of the social organization spectrum, **social dominance orders,** or hierarchies, occur when individuals are defending or possessing a "personal space" or area; this often results in the establishment of a "pecking order" or rules of rank and privilege within the population
   d. Any social system will result in one or more of the following: allocation of areas containing all required resources, defense of a point source of a particular resource, or establishment of priority of resource use
   e. In either territoriality or social dominance orders, the collection of individuals (social hierarchy group or territorial owners) influences behaviors and often determines mating success and access to requisites for survival
2. Social dominance is based on intraspecific aggressiveness and intolerance and on the dominance of one individual over another
   a. A *dominance hierarchy* is a social structure in which each individual knows its place and knows which individuals it can or cannot defeat
      (1) The top individual that wins aggressive encounters is known as the **dominant individual**
      (2) Other members of the social hierarchy are known as **subordinates**
      (3) The social structure or hierarchy in its simplest form is a "social ladder" on which the dominant individual is at the top and the subordinates are located on the lower rungs

(4) The social ladder is established based on the outcomes of aggressive encounters; winners move up the ladder, and losers move down
    b. Two opposing forces occur simultaneously in dominance hierarchies: mutual attraction among members of the species, and social intolerance, perhaps based on competition for resources
    c. Mixed-gender groups may have separate hierarchies for each sex
    d. The initial establishment of a hierarchy will involve considerable preliminary fighting
    e. Once established, the stable relationships tend to decrease aggression and fighting, because the outcome of an aggressive encounter (winner vs. loser) can be predicted from prior experience
    f. An individual's rank usually is determined at an early age by brief periods of fighting
    g. Once social structures are well established, newcomers and subordinate individuals can rise in rank only with great difficulty
    h. Once rank is attained or determined, it seldom is contested; when it is contested, the contest usually is highly ritualistic
    i. Resources often are unequally allocated such that in times of high density or scarcity, some of the lower-ranking individuals may be denied access (that is, forced to wait until all others have fed, to take leftovers, to face starvation, or to leave the group), which results in a density-dependent action that could possibly lead to population control
    j. Social dominance can influence population regulation if it affects reproduction and survival in a density-dependent way; for example, wolf packs demonstrate reproductive control within the pack in a density-dependent fashion, and reproduction is undertaken only by the dominant male and female
    k. Members of a social hierarchy often display xenophobia (fear of strangers) and generally react negatively to newcomers
    l. The dominant animals in a hierarchy have numerous advantages, such as increased availability of and access to food, greater opportunity for reproduction, choice of better nesting sites, decreased stress, and higher survival rate
    m. There also may be some compensation or consolation for the subordinate members of the hierarchy, such as improved survival as a group member (as compared to solitary life) and an outlet for the release of aggression; if group members are relatives, there is the chance to improve **survivorship** of kin and possibly the future chance of moving up in the hierarchy as dominant members are removed
3. Territoriality is the possession of a territory or defended area, more or less fixed and exclusive, maintained by an individual or by a social group (for example, a wolf pack)
    a. A territory generally is a fixed area that may shift over time and is actively defended by its owner against intrusion; territories are nonoverlapping, with sometimes dire consequences (such as death) for the intruder
    b. Territoriality is found mainly among fish, birds, and mammals, but it also may be found in some reptiles and amphibians
    c. Possession of a territory generally occurs only when the resource (for example, food, nesting sites) is predictable, dispersed over an area, or persistent in time

d. Territoriality is a form of interference competition and thereby requires simultaneous use of a scarce resource
e. Territory holders use aggressive behavior to exclude intruders and gain access to contested resources
f. Territory holders usually are adults and may have prior experience with the territory, particularly if good-quality sites are rare
g. In order for a species to maintain a territory, appropriate benefits of possessing the territory (access to resources) must outweigh the costs of defending the territory
   (1) It is reasonable to expect that there will be a balance between costs and benefits, which will result in an optimal size for a territory
   (2) If food availability is the major benefit of territory possession, then increased accrual of benefits to the territory holder would be expected as the size of the territory is increased
   (3) This benefit of increased food availability is likely to level off as the food supply increases toward a level of sufficiency or superabundance for the territory owner
   (4) As a territory's size increases, the costs of defending it are likely to increase because there is more area to patrol and greater distances over which intruders must be chased in order to repel them; furthermore, the presence of more abundant resources on a territory may serve as a greater incentive to intruders, who may be more persistent in their attempts to gain access to the territory
   (5) The size of a territory is characteristic of a species and influenced by the distribution, value, and quality of the resource
h. Characteristics of the species and of the resource itself influence the defensibility of a territory and feasibility of possessing a territory
   (1) Territorial species tend to be highly mobile in order to detect and repel intruders
   (2) The resource must be worth defending — that is, it must be predictable, long-lasting, and of sufficient quality
i. Territoriality may function to limit population size by parceling out breeding sites (or other resources) if the excluded individuals have reduced reproduction or survival
   (1) In his work with the bird species the great tit *(Parus major),* J. Krebs demonstrated that those individuals with territories in wooded areas (the preferred habitat) had greater mating success than those forced to nest in a less favorable habitat, such as hedgerows
   (2) S. Riechert (1981) demonstrated that the mortality rates of certain species of desert spiders without territories were significantly higher than those for members of the same species with territories
j. Once an individual has asserted control of a territory, intruders often avoid the territory altogether or attempt incursion onto the territory merely to test the owner
k. The territorial owner most often wins in the challenge for possession of a territory, for a number of reasons
   (1) The threat displays of territorial species have evolved properties that indicate something about the size and actual fighting capacity of the individuals involved; thus the brilliance of coloration, size of antlers, or

other characteristics used in display often are related to the ability of the individual to fight

(2) An attempt by a weaker or smaller animal to supplant a stronger one is far more likely to lead to exhaustion, injury, or death than to success; thus, a relatively weaker individual often would benefit from conserving time and energy (and also reduce the risk of injury) for later attempts to acquire a territory when the owner is older, more vulnerable, or deceased

l. Territorial battles are likely to be ritualized and involve bluffing; however, where the only chance of securing a vital resource, such as a mate or a territory, is concerned, there is likely to be actual battle and the fighting may result in death of one of the contestants

(1) The ritualized behavior is preferable to actual fighting, which carries a high risk of injury

(2) In some cases, the current or immediate disadvantage of retreat may be offset by the chances of future contests for the territory in which the loser may ultimately win

m. Territory residents or holders generally win in territorial disputes against intruders

(1) Territory holders have had a chance to assess the territory worth or value and therefore have more at stake in the contest than the intruder, who may not be able to assess the value of the territory and thus may not be as willing to fight

(2) The resident previously has invested time and energy in defending or winning the territory and therefore is likely to be more willing to rigorously defend it

(3) The intruder may be under time and energy constraints to find a territory and therefore may not be willing to escalate a battle with a territory owner, instead finding it better to retreat and try elsewhere or at a later time

n. **Leks** are specialized breeding territories that generally do not contain valuable resources but are display arenas in which mating occurs as females enter areas possessed by males

(1) Leks are used by some species of birds, antelope, dragonflies, and frogs

(2) Often, the most reproductively successful males are the ones in the center leks, where competition is most intense

(3) Obtaining and retaining a lek requires substantial expenditure of energy by the owner in defending the area, and exposes the lek owner to risk of predation

(4) Because of the intensity of competition and the risk of predation, individuals that successfully defend a lek must be healthy and strong

(5) Females choose lek owners as mates; consequently, possession of leks is important for reproductive success and provides a way by which females can discriminate among potential mates

4. A **home range** is defined as an area in which an organism normally lives; it can be considered the normal living space of a group or an individual (social vs. solitary organisms)

a. Home ranges are characteristic of most vertebrates

b. Possessing a home range carries many advantages, including familiarity with the landscape, capability of monitoring food resources, knowledge of hiding places, and recognition of neighbors; the result is reduced time and energy spent on investigating strangers and expressing aggression
c. A home range generally is not defended against intruders (if it is defended, then it is a territory)

## III. Interspecific Competition

### A. General information
1. Interspecific competition is manifested in mobile animals as aggressive encounters and often involves contest and interference competition
2. In plants and immobile animals that are fixed in space, interspecific competition is influenced by proximity to neighboring species and occurs by consuming resources in limited supply, by modifying the microclimate, or by producing toxins
3. Unlike intraspecific competition, interspecific competition often is asymmetric — that is, the consequences are not the same for both species; one usually is influenced more than the other

### B. Classical competition theory
1. The Lotka-Volterra model of competition uses mathematical equations based on the logistic growth curve to describe the relationship between two species utilizing the same resource
    a. The mathematical equations for the Lotka-Volterra model were formulated in the early 1900s by two mathematicians, Alfred Lotka of the United States and Vittorio Volterra of Italy
    b. They altered the equations for logistic population growth by adding a constant to account for the interference of one species with the growth of another; this constant essentially converts the number of individuals of one species population into the equivalent units of the other (see *Lotka-Volterra Competition Equations*)
    c. Some major assumptions are made by the Lotka-Volterra model
        (1) The environment in which the two species are located is considered to be homogeneous and stable or constant
        (2) Migration in or out of either population is negligible
        (3) Competition is the only important biological interaction taking place
        (4) If these assumptions are violated or not true, the ability of the Lotka-Volterra model to predict the outcome of interspecific competition will be adversely effected
        (5) Although the mathematics required to do so is beyond the scope of this text, it is possible to show that the Lotka-Volterra equations predict that the outcome of long-term competition between species with significant overlapping requirements is extinction of one species and total victory by the remaining species; in other words, coexistence is not possible if there is strong competition between the two species
        (6) This conclusion drawn from the Lotka-Volterra equations suggests that species existing in nature have survived competitions in their history because they have avoided them; it also suggests that speciation (the origin of new species through adaptation and evolution) is partially a

# Interspecific Competition

---

## Lotka-Volterra Competition Equations

Lotka-Volterra equations are mathematical representations of competition between two species utilizing the same resource.

Species 1:
$$\frac{dN_1}{dt} = r_1 N_1 \left( \frac{K_1 - N_1 - \alpha N_2}{K_1} \right)$$

Species 2:
$$\frac{dN_2}{dt} = r_2 N_2 \left( \frac{K_2 - N_2 - \beta N_1}{K_2} \right)$$

Where:

$r_1$ and $r_2$ equal the rates of increase for species 1 and 2, respectively

$K_1$ and $K_2$ equal the carrying capacity or equilibrium population size for each species in the absence of the other

$\alpha$ equals a measure of the inhibitory effect of one species 2 ($N_2$) individual on the population growth of species 1 ($N_1$)

$\beta$ equals a measure of the inhibitory effect of one species 1 ($N_1$) individual on the population growth of species 2 ($N_2$)

$N_1$ and $N_2$ equal the number of individuals of species 1 and 2, respectively.

**Note:** The terms $dN_1/dt$ and $dN_2/dt$ mathematically represent the change (growth or decline) of populations 1 and 2, respectively, over some time period (dt can represent any meaningful time period for a species: weeks, years, hours, minutes).

---

process of avoiding competition by promoting differences among populations
2. The essence of interspecific competition as determined by the Lotka-Volterra model is that individuals of one species suffer a reduction in reproductive success, survivorship, or growth as a result of resource exploitation or interference by individuals of another species
   a. Experiments conducted by G.F. Gause (1934) are considered classical tests of the hypotheses predicted by the Lotka-Volterra model — namely, that strongly competing species cannot coexist indefinitely and, where populations do share resources, their competition is muted and weak
   b. Gause conducted a series of laboratory experiments using the protozoan *Paramecium*
   c. He grew different species of *Paramecium* singularly and in combination in order to determine the effect of competition on the population size and persistence of each species
   d. Gause found that when *Paramecium aurelia* and *P. caudatum* were grown together in test tubes, *P. aurelia* always persisted and *P. caudatum* always died out
      (1) These results were consistent with the Lotka-Volterra predictions because both species in the experiment competed for the same food source

(2) *P. caudatum* is a relatively large and slow-growing species, whereas *P. aurelia* is smaller and reproduces more quickly — the result being that *P. aurelia* is a better competitor
        e. In another set of experiments *P. aurelia* and *P. bursaria* were grown together and neither species became extinct, although both populations leveled off at approximately half the number they would reach when growing alone
            (1) Further observation indicated that, when the two species coexisted, they were separated in space; *P. aurelia* was located at the top of the test tube, and *P. bursaria* was concentrated at the bottom
            (2) By separating themselves within the test tube, the two species were avoiding (or at least minimizing) competition and permitting their coexistence
            (3) This condition also is predicted by the Lotka-Volterra equations, because competition is substantially avoided
    3. The Lotka-Volterra equations predict that if one species in a competitive situation can grow rapidly enough to prevent the other competitor from increasing, it can reduce that population to extinction or exclude it from a particular area
    4. As a result of Lotka-Volterra predictions and experimental work, the **competitive exclusion principle** was formulated; this principle states that no two species can make the same, simultaneous demands on the environment or resources and persist together
    5. Two or more species may compete for some essential resource without being complete competitors and thus subject to competitive exclusion
    6. If overlap of resource use and competition for some resource occurs, there must be some escape mechanism — that is, some way the competitors exploit other resources differently from one another where overlap does not occur
    7. *Allelopathy* is a type of interference competition among plants in which a chemical substance is released by one species that inhibits the growth of other species
        a. Allelopathic chemicals include acids, bases, and simple organic compounds that reduce competition for nutrients, light, and space by inhibiting growth or survival of competitors
        b. Examples of allelopathy include broom sedge *(Andropogon virginicus)*, which produces chemicals that inhibit the invasion of field communities by some shrubs and trees; goldenrod *(Solidago spp.)* and asters *(Aster spp.)*, which can inhibit tree growth in certain communities; and black walnut trees *(Juglans nigra)*, which produce chemicals that inhibit the growth of many plants

## C. Niche theory
1. The *niche* of a species has been variously described as a theoretical space in which an organism lives and reproduces; an organism's place in the biotic environment; its relation to food and enemies; a specific set of capabilities for extracting resources, for surviving hazards, and for competing, coupled with corresponding sets of needs; a set of ecological conditions under which a species can exploit a source of energy to reproduce and colonize further sets of conditions; and a description of all the resources required for survival
2. The concept of the niche is closely associated with interspecific competition; the interrelationship between species greatly influences the niche
3. A niche may be considered fundamental or realized

a. The **fundamental niche** (also known as the idealized, or theoretical, niche) considers the many variables or requisites for survival (for example, food size or availability, foraging location, humidity, photoperiod, temperature range, and mineral and vitamin needs) in the absence of competitors
   b. The **realized niche** comprises those conditions under which a species actually exists and includes the impacts of competitor species
4. Niche overlap occurs when two or more species use the same resource simultaneously
   a. If the resource is in short supply, the simultaneous demand may result in competition; thus, niche overlap may be related to (proportional to) the degree of competition (that is, the more overlap, the greater the competition)
   b. Niche overlap may signify competition, but the observation of overlap does not necessarily mean that competition is occurring
5. **Niche width,** breadth, or size is a graphic representation of the range of resources used by an organism that often is determined by measuring some trait (such as bill size, food size, habitat space, temperature range, moisture range, or pH)
   a. Niche widths usually are described as narrow or broad
   b. The broader the niche, the more generalized the species
   c. Narrower niche widths may be suggestive of specialization on a particular resource
   d. Most species with broad niches possess the ability to survive within a wide range of resources or environmental conditions
6. As a result of environmental influences and interactions with competitors, the niche width of a species may change over time
   a. The niche of a species may undergo compression or constriction when the species is forced to confine feeding or other behaviors to certain portions of the niche space as a result of intense competition
   b. **Ecological release** may occur after the removal of a competitor or decline in interspecific competition along a resource gradient; this results in an expansion of niche breadth along that particular resource base to include space not formerly available to the organism
   c. A change in behavior or feeding pattern to reduce interspecific competition is termed a **niche shift**

## Study Activities

1. Compare and contrast the following terms: interspecific and intraspecific competition, contest and scramble competition, and realized and fundamental niches.
2. Construct a table listing the major types of interactions between species and whether the interaction is neutral or beneficial for one or both species.
3. Describe the benefits of a social hierarchy for the dominant individual as well as the subordinates.
4. Explain why a territory owner is most often successful at repelling an intruder.
5. Explain the ecological significance of the competitive exclusion principle.
6. Discuss how territoriality and dominance hierarchies control population size.
7. Describe the function of a lek.
8. Compare and contrast a species' territory and home range.
9. Describe how plants might respond to competition.

# 8

# Predation

## Objectives

After studying this chapter, the reader should be able to:
- Explain the various types of predation found in nature.
- Differentiate between the functional and numerical responses of predators.
- Describe the foraging behavior of a predator in terms of obtaining an optimal diet and foraging efficiency.
- Explain the relationship between plants and herbivores with respect to the defense measures and impacts of each group on the other.
- Explain some of the mechanisms used by prey species to avoid capture.
- Differentiate among the three types of mimicry.
- Describe mechanisms used by predators to improve hunting and capture success.
- Explain the role of predator-prey cycles in the regulation of population density.
- Explain the importance of cannibalism as an ecological event.

## I. Actions of Predators

### A. General information
1. A *predator* is an organism that uses other live organisms as an energy source and in doing so removes those individuals from the population
2. Prey organisms are those that serve as a source of food for other organisms; usually the prey item is consumed as food
3. Predation includes various interactions, including *herbivory* (in which the prey are plants or plant parts), *carnivory* (in which one animal preys on another), **cannibalism** (in which the predator and prey are of the same species), and **parasitoidism** (in which one organism attacks the host by laying its eggs in or on the host; the eggs hatch and the larvae feed on the host until it dies)
4. Predators often remove prey in nonrandom fashion
   a. Predators may "cull" surplus individuals from otherwise density-regulated populations
   b. Predators may preferentially take prey in marginal prey habitats
   c. In territorial prey species, predators generally take nonterritorial individuals found in suboptimal habitats
   d. Predators maximize their success by selecting those individuals from the prey population that have the least chance of escape and are easiest to catch, such as the young, the old, and the weak

## Functional Response of Predators to Prey Density

The functional response of predators can be classified as type I, II, or III. Type I predators take more prey as prey density increases, up to the point of satiation. Type II predators take an increased number of prey as prey density increases, but the rate at which they take prey declines because of handling time. Type III predators use a series of prey species in direct relation to their abundance, switching between the prey items as necessary.

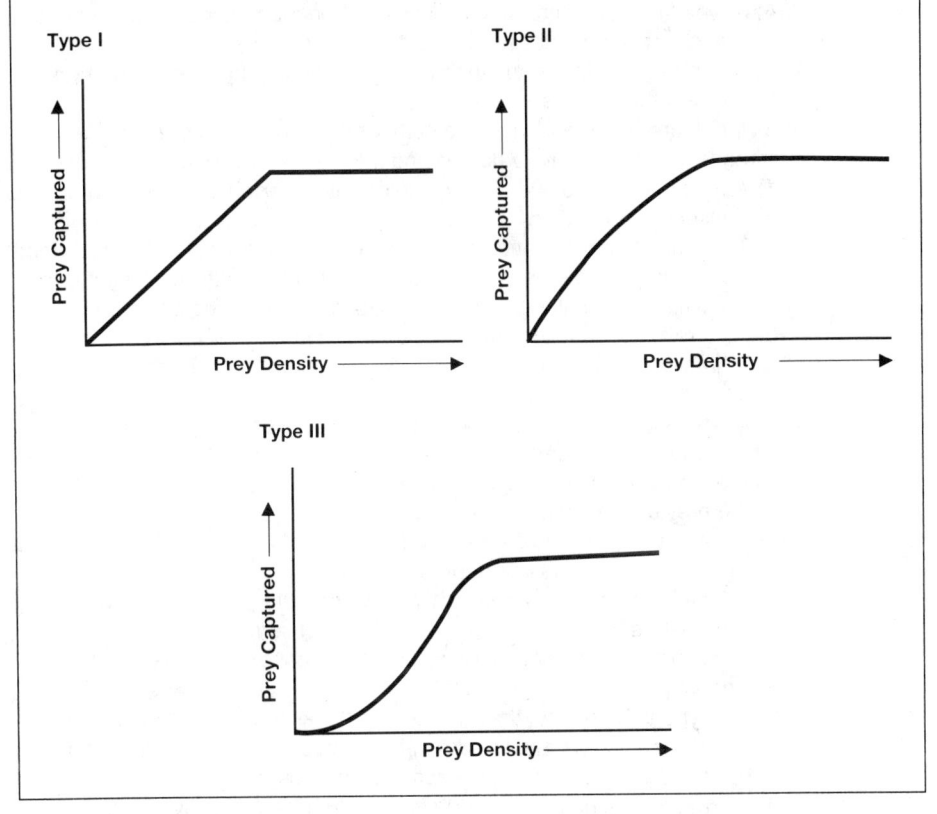

**Type I**

*Prey Captured vs. Prey Density: linear increase then plateau*

**Type II**

*Prey Captured vs. Prey Density: decelerating curve to plateau*

**Type III**

*Prey Captured vs. Prey Density: sigmoidal curve to plateau*

      e. In nonterritorial prey species, there is a greater chance of reproductive individuals being captured; these individuals may not be able to run away because of parental duties or may attempt to defend their young
   5. Predation represents a transfer of energy from one trophic level to another and often is a complex interaction between two or more species

**B. Functional response of predators**
   1. As prey density increases, each individual of the predator population may take more prey or take prey sooner; this is known as a **functional response**
   2. Important characteristics affecting the predator's functional response include the caloric value (energy content) of the prey items, prey availability, and prey attractiveness (that is, is the prey item easy to take or large enough to take?)
   3. Ecologists recognize three generalized types of functional responses, designated types I, II, and III (see *Functional Response of Predators to Prey Density*)
      a. In a type I functional response, predators take a fixed number of prey during the time they are in contact with the prey

(1) The total number of prey taken per predator per unit of time increases with increasing prey density
(2) Predation ends when the predators are satiated
(3) Examples of a type I response include trout feeding on an evening hatch of aquatic insects, and clams feeding on algae suspended in the surrounding water
(4) A type I response produces density-independent mortality of the prey up to satiation because predators at any given abundance take a fixed number of prey during the time they are in contact, usually enough to satiate themselves

b. In a type II functional response, the number of prey a predator takes increases at a progressively decreasing rate as the prey density increases
   (1) A type II response often is associated with invertebrate feeders, such as insects and spiders
   (2) The time required to pursue, subdue, eat, and digest the prey — termed **handling time** — is an important constraint or limitation on the rate at which a predator can take and subsequently process prey items
   (3) As a result of handling time, the rise in the number of prey taken per unit time decelerates to a plateau while the number of prey still are increasing
   (4) A type II functional response does not remove prey in a density-dependent fashion and so cannot control prey population size

c. In a type III functional response, the predators typically learn to concentrate on the prey as it becomes more abundant
   (1) A type III response generally is characteristic of vertebrate predators but may occur in some invertebrate predators as well
   (2) A type III response usually involves two or more prey species, with predators taking prey that are in excess of some minimal number; switching from prey to prey occurs in response to changes in prey density
   (3) In a type III response, the number of prey taken per predator increases with increasing prey density and then levels off to where the ratio of prey taken to prey available actually declines
   (4) Because the type III predator responds to prey density in a density-dependent manner, this type of response may control prey population size
   (5) In the presence of several prey species, the predator can distribute consumption among the prey in response to the relative density of the prey species
   (6) Predators may take nearly all of a prey in excess of some minimum number, referred to as **threshold of security**
      (a) Threshold of security may be determined by the availability of prey hiding places and the prey social behavior
      (b) At prey densities below the threshold of security, the predator no longer finds it profitable to hunt the prey and will switch to another prey item
      (c) In switching from one prey to another, the predator will disproportionately concentrate on one species and pay little attention to the rarer species

(7) Type III predators use a series of prey species in direct relation to their abundance, switching to the more abundant prey species and thus allowing the previously preferred species to rebuild population sizes
(8) Type III predators are said to acquire a **search image,** which is a perceptual ability to detect a cryptic prey item; once a predator locates a palatable prey item, it finds it progressively easier to locate others of the same kind
  (a) In addition to the specialization on a particular prey species, the concept of the search image includes specialization on the locale most likely to contain the targeted species, which will increase efficiency of capture
  (b) Once a search image is established, the predator has a certain inertia; it will continue to pursue the targeted prey below a certain density level at which one might predict it might otherwise shift if it were taking prey only as encountered

## C. Numerical response of predators
1. In addition to functional responses, predators may exhibit numerical changes to changes in prey density; such changes are termed **numerical responses**
2. The size of the predator population can change in response to a change in food availability (changes in prey density) by migration of predators in or out of an area or by changes in birth and death rates of the predator population that are adjusted to the prey density
3. As a result of prey density, the predator's nutritional status may be affected, which in turn affects reproduction, starvation, and overall survival
4. Time lags or delays in the numerical response are likely because it takes time for predators to reproduce or for migration to occur

## D. Foraging theory
1. Foraging theory involves the study of predator behavior to understand how a predator determines where and when it is profitable to hunt
2. Because it is unprofitable for a predator to spend time where prey density is low, predators must discover the most productive way to allocate their hunting time among the different areas and different prey species
3. **Optimal foraging** is a strategy that obtains the maximum rate of net energy gain for the predator
4. Optimal foraging involves an optimal diet, which includes the most efficient size of prey for both handling and net energy return
5. An optimal diet should include those prey items that provide the greatest energy return and should ignore less profitable items
6. Optimal foraging also includes **foraging efficiency,** which involves the concentration of activity in the most profitable areas when prey are abundant and the abandonment of areas where foraging is less profitable
7. Predators will attempt to maximize the net energy they derive from hunting, pursuing, and consuming prey and thus will attempt to maximize the following equation: $E = E_{prey} - E_{search} - E_{pursuit}$, where E equals the net energy derived by the predator, $E_{prey}$ equals the gross energy contained within the prey, $E_{search}$ equals the energetic cost of searching for the prey, and $E_{pursuit}$ equals the energetic cost of pursuit and capture

8. It often is most "profitable" to specialize on the most common prey item; for example, trout specialize on hatching insects, with the type varying depending on which kind are hatching in greatest abundance
   a. As prey species density declines, the relative efficiency of the predator specializing on that prey also is likely to decline
   b. If a predator can realize a net gain in energy by searching, pursuing, and consuming a unit of given prey, it will do so unless some other prey species offers a better return in terms of net energy; for example, trout will feed on hatching insects when plentiful but may switch to other species (or even resort to cannibalism) if insects are less available or if a single larger fish is a better item of food than many small insects
9. Handling time is an important constraint for some predators and influences their selection of prey items
   a. Although small prey often are easier to subdue than larger items, they offer less food per item
   b. Larger prey items may be rarer but offer a large food intake per item
   c. As a result of these differences, larger predators are more universal in dietary requirements than are small predators
10. Search, or foraging, time is another important component of effective and efficient foraging behavior and can be a major component of the costs of foraging by a predator
    a. Search time for small and often hard-to-find prey items can make foraging time a significant consideration; for example, birds such as wrens may spend approximately 50% of their active time foraging for food during the time of maximum insect hatching and nearly 100% of their active time foraging during winter months, when food is scarce
    b. Conversely, if the food source is predictable and easily located, such as is the case with nectar used by hummingbirds, foraging time is much lower, ranging from 5% to 10% of active time (this includes time for insect foraging)

## II. Plant-Herbivore Systems

### A. General information
1. In plant-herbivore systems, plants are the prey and herbivores are the predators
2. Herbivores may not consume the entire plant or kill it, as is usually the case in carnivory

### B. Effects of predation on plants
1. The feeding of herbivores on plants often involves *defoliation* — the destruction or removal of leaves — and consumption of fruits and seeds
2. Removal of leaves or other biomass may affect a plant's ability to survive and reproduce
   a. If herbivores remove only a part of the plant, the plant's survival depends on the amount removed and whether further removal will occur
   b. In some cases, plants are able to regenerate parts removed by herbivores
   c. Defoliation can affect the plant indirectly by weakening it and making it more susceptible to disease or by reducing its overall vigor, thereby decreasing its chances of survival

d. Seed destruction can have a direct, negative impact on **fitness** — the number of offspring that survive in the next generation
3. Some plants respond to defoliation by producing new growth and therefore drain nutrients from reserves that would otherwise be used for growth and reproduction
4. Under some circumstances, moderate grazing by herbivores can have a stimulating effect, increasing biomass production
   a. The degree of stimulation depends on the plant, its nutrient supply, and moisture conditions
   b. The stimulation may be induced by removal of older tissue, which often is less photosynthetically efficient than younger tissue
   c. Moderate grazing may stimulate biomass production in a deciduous forest canopy
   d. In general, grasses increase biomass production when grazed, but only up to a point, after which production declines

**C. Herbivore nutrition**
1. Plant tissue quality is important to herbivores because of the complex digestive processes required to break down plant cellulose and convert plant tissue into animal tissue
   a. The nitrogen content of plant tissue must be high for herbivores to survive
   b. High-quality plant tissues include young, soft, and green plant parts as well as storage organs, such as roots, tubers, and seeds
2. Plants produce a variety of metabolic products known as secondary plant substances, which can affect the health and reproduction of herbivores

**D. Plant defense**
1. Throughout evolutionary history, plants have attempted to develop defense measures against herbivores
2. Plants produce a number of structural defenses that make penetration by predators difficult; these include tough leaves, hard seed coats, thorns, thick bark, and spines
3. By synchronizing reproduction to produce a maximum number of seeds in a very short period, plants may "flood the market," so that the predators may be satiated and a certain percentage of the seeds will escape; this synchronizing of seed or fruit production may be within one species or among several species
4. Plants produce various chemical defenses in the form of chemical compounds known as secondary plant products; these include phenolics, alkaloids, tannins, and mustard oils
   a. These secondary substances can induce a hormonal imbalance in some grazing herbivores that results in infertility and reduced milk production
   b. Other substances produced by plants serve as repellents or chemically bind with nutrients to make them unavailable to the herbivore
   c. The production and storage of these chemicals is energetically expensive for the plant and seems to require a trade-off between defense and reproductive effort
   d. Although plants possess powerful chemicals for defense, specialized herbivores have evolved countermeasures
      (1) The major detoxifying system found in animals is a series of enzymes referred to as *mixed function oxidase*

(2) Mixed function oxidase activity is found in all animals; it is located in the liver tissue in vertebrate animals and in the gut, fat bodies, and Malpighian tubules in insects
(3) Mixed function oxidase activity converts foreign, toxic chemicals to less dangerous chemicals that can be eliminated from the body by the excretory system
(4) The mixed function oxidase system is able to detoxify a broad range of toxic compounds that may be encountered by the animal

## III. Herbivore-Carnivore Systems

### A. General information
1. Carnivores are not faced with a lack of quality in their food; there generally is sufficient protein content in prey animal flesh
2. Carnivores are faced with the problem of obtaining sufficient quantity of prey

### B. Prey defense
1. Prey species may use various mechanisms to defend themselves against predation; these include behavioral mechanisms, chemical defenses, warning coloration, mimicry, cryptic coloration, armor, and predator satiation
2. A prey species may respond to the presence of a predator in a variety of ways
    a. The prey species may choose to ignore the predator if attack is unlikely and the hunger level of the prey species is high; it may be advantageous for the prey to continue to eat
    b. The prey species may move out of contact with the predator and hide or sneak away
    c. The prey may freeze in position upon detecting a predator; this is useful in situations where the predator is moving in a direction such that its search path will not include the prey in direct sight or line, or it may be advantageous if there is a short prey-predator distance (no chance of fleeing to escape) and the prey is less conspicuous to the predator or difficult to detect
    d. The prey may flee to escape the predator; for fleeing to be a successful strategy, the prey must be able to outdistance the predator or evade capture for a period sufficient for escape
        (1) For example, zebras, gazelles, and wildebeests generally can outrun tigers or lions at distances of 20 to 35 meters or more; if the predator-prey distance is less, the lions usually will make a kill
        (2) Zebras, gazelles, and wildebeests can outrun cheetahs only if they have a head start of at least 350 meters
    e. The prey may mob or confront the predator as a group in an attempt to confuse it and alert others; this strategy is common in birds
    f. By living in groups, the prey population will have more detectors (eyes, ears, noses), which might lead to earlier detection of the predator
        (1) The improvement in predator detection from grouping probably does not continue to increase with ever-increasing group size
        (2) There likely will be some optimal group size above which prey are likely to interfere with one another and actually impede escape from a predator

3. Chemical defenses are widespread among animals and demonstrate a variety of mechanisms by which animals repel potential predators
   a. When attacked by a predator, some species of fish release chemicals from their skin into surrounding water; these substances serve to alert other individuals
   b. Arthropods, amphibians, and snakes produce strong, malodorous secretions to repel predators
   c. Some mammals (for example, shrews, skunks, and ferrets) produce obnoxious secretions from specialized anal glands
4. Animals with chemical defenses often possess warning colors, bold colors, or special color patterns that serve as a warning to potential predators
   a. The black and white stripes of a skunk, the black and yellow coloration of wasps and bees, and the bright coloration of monarch butterflies are examples of warning coloration
   b. Predators must learn by prior negative experience with the prey in order to associate the warning color with unpleasantness or pain
5. Some prey species have evolved characteristics (for example, color, shape, markings) that match another species in order to mislead predators and either gain an advantage or escape predation — a defense mechanism known as *mimicry*
   a. In mimicry, one species known as the *mimic* bears a resemblance to another species, known as the *model*
   b. The model possess the defense mechanism (usually chemical) or other advantage; the mimic gains protection or advantage by closely resembling the model
   c. In *Batesian mimicry,* a palatable mimic resembles an unpalatable model; numerous examples of this mechanism are seen in nature
      (1) Viceroy butterflies *(Basilarchia archippus)* are a palatable species that mimics the toxic Monarch butterflies *(Danaus plexippus)* in order to escape predation by birds
      (2) The palatable and harmless South African desert lizard *(Heliobolus lugubris)* mimics the color pattern and posture of an unpalatable beetle *(Anthia)* that sprays a noxious fluid when attacked by a predator
      (3) Many harmless snakes mimic the conspicuous red, white, and black markings of the poisonous coral snake
      (4) For Batesian mimicry to be successful, the population size of the model species must be larger than that of the mimic species to ensure that the predator is more likely to encounter a member of the model species and have an unpleasant experience, which will serve to further reinforce the avoidance of both the model and mimic populations as potential prey
   d. In *Mullerian mimicry,* two or more unpalatable species superficially resemble one another
      (1) In Mullerian mimicry, all individuals of all prey populations involved are unpalatable and so they always present a distasteful experience for the predator and thus collectively avoid predation
      (2) Many wasps and stinging insects have a generalized yellow and black coloration that can be quickly identified by potential predators
6. Color patterns and accompanying behaviors, known as cryptic coloration and behaviors, have evolved to conceal prey from predators

a. These colorations, shapes, movements, and behaviors tend to make the prey less visible and thus less likely to be detected
b. Some color patterns enable the species to blend in with the natural environment; for example, green grasshoppers blend in with vegetation, and chameleons are able to adjust their body color to match their background
c. Many insects have evolved shapes to resemble aspects of their environment; examples include walkingsticks, which resemble twigs, and katydids, which resemble leaves

7. Some of the most effective antipredator defenses involve protective structures or weapons to deter predators
   a. Clams, armadillos, turtles, and many beetles withdrawal into a protective shell or other structure upon attack by a predator
   b. Porcupines, hedgehogs, and echidnas have modified hairs (quills) that are sharp and deter predators

8. In **predator satiation,** reproduction of the prey is timed so that most of the offspring are produced in a short period
   a. With large numbers of offspring produced at one time, so much prey becomes available that the predator becomes satiated and a certain percentage of the offspring can escape
   b. Wildebeest and caribou synchronize the birth of otherwise defenseless young so that large numbers of offspring are present simultaneously, thus permitting some offspring to survive because predators will fail to capture them
   c. The *Magicicada* cicadas appear at 13-year and 17-year intervals in enormous numbers and quickly satiate predators, thus allowing a sufficient number to survive

## C. Predator offense

1. Predators have evolved ways to improve hunting and capture success and to counteract the antipredator measures of prey
2. The hunting tactics of predators vary according to the type of predator and the type of prey
   a. Predators have three general methods of hunting: ambush, stalking, and pursuit
      (1) Ambush hunting, which involves lying in wait for prey items, has a low frequency of success but requires a minimum energy expenditure; it is typical of certain insects, frogs, lizards, and alligators
      (2) Stalking, which ultimately involves quick attack, generally entails a long search time and minimal pursuit; it is common in some predators, such as herons and some cats
      (3) Pursuit hunting, which involves significant chase, capture, and handling costs, is typical of highly mobile predators, such as hawks and some canines
      (4) Stalkers and ambush hunters can energetically afford to take smaller prey items; pursuit hunters generally must take larger prey in order to obtain sufficient energy
   b. A balance must be struck with regard to prey size; prey must be large enough to provide sufficient energy return, yet not so large as to be too big, difficult, or dangerous to handle

c. Predators may take advantage of other predators' success in order to improve their own energetic balance
    (1) One predator may attempt to scavenge or steal an already captured prey item
    (2) Some predators (for example, coyotes, hyenas, African wild dogs, lions, vultures, magpies) reduce hunting costs by eating carrion from previously killed prey of other predators
d. The capture of large prey often requires cooperation or pack hunting
    (1) For example, wolves prey in packs on caribou, moose, and deer
    (2) Mackerel, bluefish, and striped bass hunt in schools and herd their prey
e. Another tactic used by predators is *aggressive mimicry*, in which predators evolve shapes, color patterns, or structures to improve hunting ability
    (1) Many species of praying mantises mimic flowers to entrap insect prey
    (2) The anglerfish possesses a body appendage resembling a worm that it wiggles to lure prey within reach
    (3) Fireflies of the genus *Photouris* mimic the light-flashing patterns of females of other species to attract and prey upon the males of other species
3. Many factors influence a predator's hunting success
   a. Environmental factors (for example, wind, light, ground cover, snow depth, temperature, and moisture) can affect prey detection, selection, and hunting success by impairing a predator's vision, olfactory clues, or movement
   b. The age and experience of the predator also can affect hunting success; for example, a predator may learn from experience and therefore increase hunting success with time
   c. Hunting success also varies with the type of prey chosen; for example, coyotes make successful captures on 41% of their attempts when hunting ground squirrels but on only 19% of their attempts when hunting voles
   d. The size of a hunting pack influences hunting success and may vary depending upon the predator and the prey
      (1) When hunting wildebeests, spotted hyenas generally form packs of two to three animals; but when hunting zebras, they usually form packs of 11 to 27 animals
      (2) When hunting gazelles, lions generally are successful 15% of the time when solitary and 32% of the time when in prides of two to four
      (3) Golden jackals hunting gazelles are successful only 16% of the time when solitary but 67% of the time when in pairs
   e. Hunting success among predators varies greatly, ranging from low success rates of 5% for wolf packs hunting moose to high success rates of 80% to 90% for osprey hunting fish

## D. Regulation of predator and prey
1. In order for predators to limit prey populations, it is necessary for them to reduce the prey population by reducing the prey's biotic potential in a density-dependent manner
2. Although there is a potential for predators to limit prey population size, there is little evidence that predation accomplishes such regulation in the natural world
3. Predators generally take vulnerable prey — those that are sick, old, young, or weakened — and generally do not remove individuals in the reproductive stage of life, which tend to be healthy, strong adult individuals

a. Under most circumstances, predation will have an effect on the prey population only to the extent that young prey are taken; even then, a very large number of the young would need to be removed to have an impact
b. Predatory removal of vulnerable prey is compensated for by future reproduction in the prey population
c. When prey populations decline dramatically for reasons other than predation, predators may be forced to take reproductive stock and thus drive the prey population even lower, perhaps even to collapse and extinction
4. Assessing the role of predation in regulating prey populations is difficult because of the human intrusion that can result in the removal of entire populations of predators and prey or large-scale habitat alteration or destruction

## E. Predator-prey cycles
1. Populations of predators and prey often vary closely in what appear to be linked cycles (predator-prey cycles), although cause and effect are difficult to determine
2. A persistent predator-prey system requires a refuge in time or space (in which a segment of the prey population can escape predation), alternate prey available to the predator when the primary prey species reaches low densities, and a Type III predator capable of learning and switching prey
3. Based on observation and empirical data, most ecologists now believe that predation may stabilize prey populations in theory; however, more often, the relationship results in unstable fluctuations
4. Initial examination of the predator-prey system of lynx and snowshoe hare suggested a cycle of 10 years' duration, but more recent work has suggested a very different interpretation of the relationship
    a. Earlier interpretation of the fur-trapping returns of the Hudson Bay Company led investigators to describe a cycle of population increase and decrease of the hare and lynx
    b. As initially described, the snowshoe hare and lynx populations cycled more or less in tandem; the hare population increased as the lynx population declined, then the lynx population began to increase as the hare population increased
    c. Most ecologists now believe that the original interpretation of the cycles was based upon noncomparable data; the hare data were from the Hudson Bay region of eastern Canada and the lynx data were from western Canada
    d. More recent investigation suggests that the cycle is driven by the relationship between the hare and vegetation, and so the cycles were not connected as originally thought
    e. As the hare population increases, increased herbivory causes a decline in vegetation caused by excessive browsing
    f. The decline in food supply causes a high winter mortality of juvenile hares and lowered reproduction the following season
    g. A decline in hares results in a decline in the lynx population
    h. Thus the snowshoe hare and lynx system actually is a vegetation-hare cycle in which the hare population is controlled by the food supply and the lynx population is controlled by number of hares present

i. The major impact of the lynx appears to be suppression of hare (prey) population during periods of low density and the protraction of these low-density periods

**F. Cannibalism**
   1. Cannibalism is a special form of predation in which the predator and prey are of the same species
   2. Cannibalism occurs in a wide range of animals, including protozoans, centipedes, mites, spiders, insects, frogs, birds, and mammals
   3. Cannibalism accounts for up to 46% of the mortality in eggs and chicks of herring gulls, 8% of the mortality in young Belding ground squirrels, and 25% of the mortality in lion cubs
   4. Cannibalism is most likely to occur in stressed populations, particularly those experiencing food shortages and crowding
   5. The most commonly cannibalized individuals are those that are most vulnerable — that is, the young, the small, eggs, nestlings, and so on
   6. The advantages to the cannibals (predators) are that the predator obtains a meal, intraspecific competition is reduced, the fitness of a competitor is reduced, and a potential predator (fellow cannibal) is eliminated
   7. By reducing intraspecific competition at times of resource shortage, cannibalism may reduce the likelihood of local extinction
   8. Cannibalism can be detrimental if the survivors become too aggressive and destroy their own offspring and thus reduce their own fitness
   9. Cannibalistic individuals generally do not discriminate between members of their own species and other prey; rather, they are opportunistic, taking whatever prey item is available

## Study Activities

1. Sketch a series of graphs showing the three types of functional responses of a predator.
2. Compare and contrast Batesian, Mullerian, and aggressive mimicry.
3. List at least four behavioral mechanisms used by prey to avoid predation.
4. Explain what must occur if a predator is to control prey population density.
5. Describe the concept of predator satiation as used by plants and animals to reduce the impact of predation on population density.
6. Explain the positive and negative effects of cannibalism.

# 9

# Life History Patterns

## Objectives

After studying this chapter, the reader should be able to:
- Describe the trade-offs made by organisms with respect to reproduction, maintenance, and growth.
- Explain the concepts of parental investment and parental care.
- Describe the differences among life strategies of plants and those of animals.
- Describe various life history patterns and strategies using the concept of $r$ and $K$ selection.
- Understand the various mating systems that have evolved and the factors that give rise to them.
- Explain the concept of sexual selection.

## I. Reproductive Effort

### A. General information
1. Life history patterns include any pattern of life process or behavior that affects an organism's fitness by the efficient collection or use of resources
2. **Reproductive effort** is the sum of the current reproductive output and the residual or future reproduction; it takes into account the proportionate contribution of an individual to future generations in terms of producing offspring
3. Because the resources (for example, time and energy) at an organism's disposal are finite and limited, energies devoted to current reproduction often can lead to reduction in survivorship, growth, or future reproduction
4. An organism's life history is the result of trade-offs among reproduction (present and future), survivorship, competition, predator avoidance, growth, and maintenance

### B. Energy costs
1. To reproduce, parents must expend energy; a certain minimum amount of energy is needed to produce viable offspring
2. Theoretically, as parents apportion available energy among larger numbers of individual offspring, each offspring receives less energy; there is likely to be a point at which the number of offspring is so large that each is unlikely to receive sufficient energy for survival
3. Parents have a finite amount of energy to apportion among growth, maintenance, and reproduction; if organisms apportion a large proportion of energy to repro-

duction, they may grow more slowly to the next life stage, may fail to reproduce at a later date, or may even fail to survive
4. Patterns of energy (or any resource) allocation will have an impact on parent and offspring survival

## C. Parental investment
1. The amount of energy, time, and resources spent in the production, nurturing, and care of offspring is known as *parental investment*
2. There are two basic patterns of parental investment: to apportion parental investment among many small young, or to concentrate investment in a few larger young; both apportionment strategies attempt to maximize the survival of offspring under different conditions
    a. Producing many small offspring, each with a relatively small probability of survival, will collectively ensure that at least some will survive
    b. Producing fewer, larger offspring provides a higher probability that any one individual survives
3. Within the range of available resources, parents must adjust the number of young that they rear without exceeding their ability to provide resources or significantly reducing their own chances of survival
4. Many bird and insect species demonstrate the ability to adjust the number of offspring produced and the care given to those offspring in response to available resources or changes in available resources
    a. For example, the common grackle *(Quiscalus quiscula)* has an asynchronous hatching pattern in which some eggs hatch earlier than others
    b. If food is scarce at hatching, the parents do not feed the later-hatching chicks and thus allow them to die of starvation
    c. The asynchronous hatching combined with the selective feeding ensures survival of at least some of the young even during times of food shortage but provides an opportunity to successfully rear a larger brood when resources permit
5. Parental investment also includes the care of young from birth or hatching until independence
    a. The extent and type of care provided generally is influenced by the maturity of the offspring at birth
    b. Some birds and mammals invest considerable energy in offspring during the developmental stage (incubation or gestation); others provide parental care by investing energy in young after hatching or birth by caring for and protecting juveniles
    c. Generally birds and mammals are either altricial or precocial at hatching or birth
        (1) *Altricial* animals generally are born naked and helpless or nearly so and require considerable parental support
            (a) Altricial birds are completely helpless at hatching; their eyes are closed and they have few or no feathers
            (b) Young mice, bats, and rabbits are considered altricial because they are born blind and naked
        (2) *Precocial* animals generally are able to move about at or shortly after birth but may still require some parental support

(a) Precocial birds hatch with open eyes and completely covered with down feathers; they generally are mobile to some degree, often able to leave the nest within a couple of days
(b) Mammals are either semiprecocial (for example, wolf, fox, dog, cat, deer, moose, horse, and pig), born with hair and able to move around soon after birth, or fully precocial (for example, seal and whale), mobile immediately on birth
d. Parental care is not well developed in invertebrate species, although social insects such as ants, bees, and wasps do provide some parental care in the form of food and defense
e. Some species of fish provide parental care, whereas others provide none at all; as a general rule, the larger the size of the egg relative to the size of the parent, the greater the amount of parental care provided
   (1) Large-mouth bass and catfish produce relatively large eggs and actively protect their eggs and young
   (2) Trout lay large numbers of small eggs and provide no parental care

## D. Reproductive value
1. *Reproductive value* is the potential reproductive contribution of an individual at a particular age relative to that of a newborn individual at the same time
2. The reproductive value of a newborn individual is influenced by the state of the population in which it is found
   a. In an expanding population, a newborn's reproductive value is low, because the probability of dying before reproducing increases and because the future breeding population to which the newborn will belong is larger, thus diminishing the individual contribution to the overall gene pool
   b. A newborn entering a declining population will contribute more to future generations than present progeny and so is worth more and has a larger reproductive value
3. Reproductive values generally are easier to calculate for females than males because of the difficulty in most species in determining the number of offspring sired by a male

## E. Reproductive costs
1. To contribute the greatest or maximum fitness to future generations, organisms need to balance short-term benefits of reproduction with long-term costs — in other words, balance the production of present and future progeny
2. The balance between short-term and long-term reproduction will be affected by the environmental conditions prevailing at the time of present and future reproduction, the health of reproductive individuals, and population size; the last will influence competition and predation
3. The trade-off between short-term and long-term reproduction also will be influenced by the impact of current breeding on an individual's future survival and ability to reproduce

## II. Reproductive Patterns

**A. General information**
1. Because organisms have access to finite amounts of resources and have finite life spans, various patterns of reproduction have evolved to optimize reproductive value
2. Plants and animals can be categorized based on their general reproductive patterns
   a. Animals generally are classified as semelparous or iteroparous, based on the frequency of reproduction
   b. Plants typically are classified as annual, biennial, or perennial, based on their life cycle or life span

**B. Semelparity and iteroparity**
1. Two broad categories of reproduction modalities, known as semelparous and iteroparous, result from the trade-off between present and future reproduction
2. *Semelparous* organisms have only one major reproductive effort in their lifetime; they invest all their energy in growth, development, and storage and then expend all that energy in one large, often suicidal, reproductive event
   a. Semelparity is common in most insects and invertebrates, as well as some fish (for example, salmon)
   b. Many semelparous animals are small and short-lived and occupy ephemeral or disturbed habitats where chances of future parental investment are low and the best strategy appears to be expending all their energy in one bout of reproduction
   c. Other semelparous species are long-lived and delay reproduction until the end of their lifetime; examples include salmon and periodical cicadas
3. *Iteroparous* organisms produce fewer young at one time and repeat reproductive efforts throughout their lifetime
   a. Iteroparous organisms face the dilemma of early versus later reproduction during their lifetimes
      (1) Because these organisms operate within a finite resource budget, early reproduction may reduce survivorship and the potential for later reproduction, whereas later reproduction increases growth and improves survivorship but reduces **fecundity**
      (2) Iteroparous species must strive for a balance between early and late reproductive effort to obtain optimal lifetime reproductive success
   b. Most vertebrates (that is, mammals, birds, reptiles, amphibians, and fish) are iteroparous

**C. Annual, biennial, and perennial plants**
1. *Annual plants* complete their life span in 12 months or less; every individual breeds during one particular season, then dies before that same season the next year
   a. In annuals, each generation is distinguishable from others and is thus considered discrete; the only overlap of generations is with adults and offspring immediately after the reproductive season (young seedlings that have germinated and seeds still in the ground that have not yet germinated)
   b. Desert annual plants have a substantial seed bank, with germination occurring on occasions when conditions are favorable (for example, immediately following precipitation)

c. Many weedy plants, wildflowers, and food crops are annuals that survive winter or other harsh environmental conditions as seeds
2. *Biennials* have a life-span of 2 years, with reproduction (flowering) usually occurring during the second year after a first year of vegetative growth; examples include beets, carrots, radishes, and their wild relatives such as mustards and Queen Anne's Lace
3. *Perennials* (for example, trees, shrubs, and some grasses) live many years and commonly reproduce repeatedly during their lifetime
4. Although plant life cycles often can be easily categorized as annual, biennial, or perennial, the various life cycles may merge into more complex ones without any sharp discontinuity; it often is difficult or unnecessary to make sharp distinctions

## III. *r* and *K* Selection

### A. General information
1. The concept of *r* and *K* selection originated with R. MacArthur and E.O. Wilson (1967) and was later elaborated on by E. Pianka (1970) in their study of the different types of selection pressures on plants and animals colonizing islands
    a. MacArthur, Wilson, and Pianka observed that environments present a range of selective pressures to which organisms must adapt in order to survive and successfully reproduce
    b. In some cases, the most successful strategy may be to devote significant resources to reproduction, ensuring representation in the next generation; in other cases, a more successful strategy might involve the diversion of resources to competition or predator avoidance, ensuring successful survival to reproductive age
2. Uncrowded or empty environments favor colonists with the ability for rapid population growth; as the environment becomes crowded it becomes necessary to divert resources to competition, thereby reducing reproductive effort
3. Species living in harsh or unpredictable environments often allocate more energy to reproduction than to growth and maintenance; these are known as *r*-selected
    a. Allocating a large proportion of resources to reproduction makes sense because the harshness or unpredictability of the environment in which *r*-selected species find themselves is likely to be fatal to large numbers of individuals
    b. The *r* of *r*-selected denotes a large biotic potential or intrinsic rate of natural increase
4. Species living in more stable environments commonly allocate more resources to nonreproductive activities; these are called *K*-selected
    a. The allocation of resources to other activities, such as competition or predator avoidance, is necessary so that these species may reach the reproductive stage
    b. The *K* of *K*-selected denotes the carrying capacity of the environment and is suggestive of the fact that the population levels of these species may be at or near the carrying capacity
5. *r* and *K* selection should be thought of as a continuum, not mutually exclusive; the *r*- and *K*-selection classification is conceptually useful, but often it is difficult to

> **Characteristics of r- and K-selected Species**
>
> This table lists the major characteristics of r- and K-selected species. Note that a particular species may have some characteristics of each type.
>
> | CHARACTERISTIC | r-SELECTED SPECIES | K-SELECTED SPECIES |
> |---|---|---|
> | Maturation time | Short | Long |
> | Life span | Short | Long |
> | Mortality rate | High, particularly among young | Generally low |
> | Number of offspring per reproductive episode | Many | Few |
> | Number of reproductive episodes per lifetime | Usually one | Many |
> | Age at first reproduction | Early | Late but repeated |
> | Size of offspring or propagule | Small | Large |
> | Parental care | Little or none | Extensive |

place naturally occurring species into one class or the other (see *Characteristics of* r- *and* K-*selected Species*)

**B. r selection**
1. r-selected species generally are short-lived, have high reproductive rates, and produce large numbers of offspring with low survival rates but rapid developmental rates; they tend to inhabit temporary or ephemeral habitats
2. Mortality of r-selected species tends to be environmentally induced and is relatively independent of population density
3. r-selected species are capable of wide dispersal and are good colonizers; they are characteristic of the early stages of ecological succession

**C. K selection**
1. K-selected species are competitive, with stable populations of relatively long-lived individuals
2. K-selected organisms often are large and produce relatively few seeds, eggs, or young
3. K-selected species inhabit environments in which mortality is density-dependent and death is a result of competition, stress, or some other factor that increases in intensity as density increases
4. K-selected animal species often exhibit an elaborate and extended period of parental care; the seeds of K-selected plants usually contain large food stores to give seedlings a strong start
5. K-selected species often are specialists and efficient users of their environment; but their populations often are at or near the carrying capacity, so the population is resource-limited
6. K-selected organisms generally are poor colonizers and often are characteristic of later stages of ecological succession

# IV. Mating Systems

## A. General information
1. The combination of behavioral mechanisms and social organization involved in obtaining a mate is referred to as a *mating system*
2. Mating systems vary in form both among and within species, involving pairs and groups of individuals
3. The relationships between males and females within a population are influenced by ecological conditions and the ability of individuals to control access to mates or resources
4. Mating systems range widely from monogamy through many variations of polygamy

## B. Monogamy
1. *Monogamy* involves the formation of a strong bond or attachment between one male and one female
2. Monogamy is most prevalent among birds and is rare among mammals, except for humans, foxes, mustelids, and beavers
3. Monogamy occurs when neither partner has the opportunity to monopolize the other either by direct control or through control of resources
4. It is most common in species in which both parents are needed for successful rearing of offspring and in which maximum fitness for both parents is achieved when both share in parental care

## C. Polygamy
1. In **polygamy,** an individual has two or more mates, none of which is mated to other individuals, and a bond exists between the polygamous individual and each mate
2. Polygamy frees one partner from parental duties, which in turn allows that individual to devote more time and energy to intraspecific competition for mates and resources
3. The more that critical resources, such as food or quality habitat, are unevenly distributed within the species' environment, the greater the opportunity for the emancipated individual to control the resources and available mates
4. The basic forms of polygamy are determined by the sex of the controlling mate
    a. In *polygyny,* the male gains control of or access to two or more females
    b. In *polyandry,* the female gains control of or access to two or more males
5. There are many variations of polygyny and polyandry, each based upon whether the controlling sexual partner directly limits access to its mate or limits access by controlling a vital resource, such as food, nesting sites, or care of young

## D. Hermaphroditism
1. *Hermaphrodites* function as both male and female in sexual reproduction by producing both sperm and eggs
2. In most hermaphroditic organisms, self-fertilization does not occur; rather, individuals mate with others to exchange genetic material and thus introduce genetic variation into the population
    a. For example, earthworms are hermaphrodites but mate with others to exchange gametes

        b. Self-fertilization among animals is very rare and appears to be restricted to a few species of land snails
    3. Many hermaphroditic plants can self-fertilize but have evolved means to prevent self-fertilization by having male and female flower parts mature at different times within a particular flower
    4. Although many plant species possess mechanisms to limit or prevent self-fertilization, sometimes self-fertilization is advantageous, such as when an individual colonizes a new habitat by reproducing itself
    5. Other hermaphroditic plant species (for example, jewelweed) self-fertilize under stressful environmental conditions when exchange of pollen with other plants has been unsuccessful; thus self-fertilization ensures the next generation
    6. The evolutionary problem with self-fertilization among hermaphrodites is the loss of genetic diversity and the accumulation and expression of deleterious genes in the population

## V. Sexual Selection

### A. General information
   1. The various mating systems tend to limit the access of individuals to mates; consequently, competition to obtain mates becomes important in a population
   2. In many cases, members of one sex compete among themselves to mate with members of the opposite sex, and members of the opposite sex show a preference for those that win
   3. Competitive mating results in differential production of offspring by different individuals, a process known as **sexual selection**
   4. Sexual selection ultimately results in the evolution of morphological and behavioral traits that influence mating success

### B. Mate selection
   1. Females often make the selection of males for mating; because females invest more in reproduction than males, it is to their advantage to be selective in choosing a mate
   2. The mechanisms or criteria used for mate selection are not completely understood
        a. The female may select a winner of combat or ritualized display; this occurs in bighorn sheep, elk, and seals, for example
        b. The female may choose a mate based on the quality of the male's territory, as in some bird species (such as in redwing blackbirds)
   3. As a result of the competition for mates and female selection of mates, certain characteristics are likely to evolve within males
        a. If females consistently choose the winners of combat, then there will be considerable selective pressure to increase male size, because larger males are more likely to win
        b. If females choose males on the basis of display patterns, then selective pressure for bolder and more colorful display patterns in males will result
   4. As a result of the continued selective pressures on one sex, over time the two sexes may diverge and look quite different
        a. For example, male deer possess antlers and females do not

  b. Also, among many songbirds, the male is brightly colored and the female often is drab
 5. This divergence of the sexes caused by sexual selection is known as *sexual dimorphism;* it often results in one sex being brightly colored or larger than the other

## Study Activities

1. List the various life-sustaining activities an organism must perform to survive, then explain why organisms must allocate energy or resources among these activities.
2. Compare and contrast altricial and precocial offspring. What are the relative advantages and disadvantages of each type?
3. Briefly explain the importance of semelparous and iteroparous reproductive patterns. What are the relative trade-offs inherent in each reproductive pattern?
4. List all the attributes of an *r*-selective species and its environment. Do the same for a *K*-selective species.
5. Explain how sexual selection often results in sexual dimorphism.
6. List all the possible mating patterns exhibited by animals; provide a short description and an example of each.

# 10

# Coevolution and Mutualism

## Objectives
After studying this chapter, the reader should be able to:
- Explain the role that coevolution plays in the development of associations among organisms.
- Describe the different forms of mutualism and how the partners in each form relate to one another.
- Discuss examples of mutualism and describe the nature of the interactions among the participants.

## I. Coevolution

**A. General information**
1. The classical definition of *coevolution* is the joint evolution of two interacting populations in which selection is reciprocal and any change in one species changes the selective forces for the other species
2. More recent views of coevolution suggest that the classical definition may be too restricted, because it implies that the interacting species were associated with one another for a long period
3. Rather than experiencing a long shared evolutionary history, the coevolved pairs probably evolved in various habitats over time and with different selection pressures; only when they invaded new environments did they come in contact with new species
4. Regardless of the evolutionary history of coevolution, the essential point is that the actions or activities of one group of species produces a generalized response in another; they become adapted to one another and thus coevolve

**B. Types of coevolution**
1. Predator-prey relationships often are considered coevolutionary relationships; under the selective pressure of predators, the prey tend to evolve toward better escape mechanisms or better cryptic defenses to avoid detection
2. Mutualism, or symbiosis, also is considered a form of coevolution because of the joint, positive, and reciprocal nature of the evolution relationship between species partners
3. Other general examples of coevolution may be found in the relationship between species of flowering plants and their pollinators

# II. Mutualism

## A. General information
1. **Mutualism** is a type of symbiosis resulting from a positive reciprocal relationship between dissimilar organisms in which both partners benefit by enhanced survival, growth, or fitness (successful reproduction)
2. Scientists believe that mutualism probably evolved from predator-prey, parasite-host, or commensal relationships that eventually became mutually beneficial

## B. Types of mutualism
1. Mutualism may be obligate or facultative, depending on whether or not the relationship is necessary for the partners' survival, and symbiotic or nonsymbiotic, depending on whether or not the partners live in very close association with one another
2. In *obligate symbiotic mutualism,* the individuals interact physically (and often intimately) and their relationship is obligatory — at least one of the partners cannot survive without the other
   a. The fungi that commonly grow in association with plant roots, known as mycorrhizae, provide an example of obligate symbiotic mutualism
      (1) In *mycorrhizal* relationships, the fungi act as extended plant roots, enhancing the roots' capability to absorb nutrients and reducing their susceptibility to pathogens
      (2) The roots of the host plant provide structural support and a constant supply of carbohydrates (transported to roots from photosynthetic leaves) for the mycorrhizae
      (3) The plant and the mycorrhizae are fully dependent on one another; any alteration that impairs one member of the association is detrimental to the other
   b. Coral reefs provide another example of obligate symbiotic mutualism
      (1) Small photosynthetic algae live within the lining of the oral cavity of the coral animals
      (2) The coral benefits though the use of photosynthetic products from the algae; in turn, the coral removes and recycles nutrients from the water that are essential to the algae's survival
3. In *obligate nonsymbiotic mutualism,* the two species live independent lives but cannot survive without each other
   a. An example of this type of mutualism is demonstrated by the yucca plant and the yucca moth
      (1) The yucca plant depends on the yucca moth for pollination, and the larvae of the yucca moth are obligate predators of yucca seeds
      (2) To ensure successful pollination, the yucca plant sacrifices a percentage of its seeds to the moth larvae
      (3) The larvae will not survive unless they are deposited as eggs and hatched in yucca seeds
      (4) During the rest of their life cycles, the moth and the plant are not associated with one another
   b. Many figs also have obligatory relationships with pollinating wasps
      (1) The wasps lay eggs in fig flowers, then the eggs hatch into larvae that consume some fig seeds
      (2) In turn, the wasps ensure pollination of the fig

## Ant-Acacia Mutualism

The coevolutionary relationship between ants and the acacia tree is an example of obligate nonsymbiotic mutualism. Nutritional nodules at the tips of the acacia's leaves are eaten by the ants (A). The tree's enlarged thorns provide housing for the ants (B). Nectaries at the base of the acacia's leaves provide additional food for the ants (C).

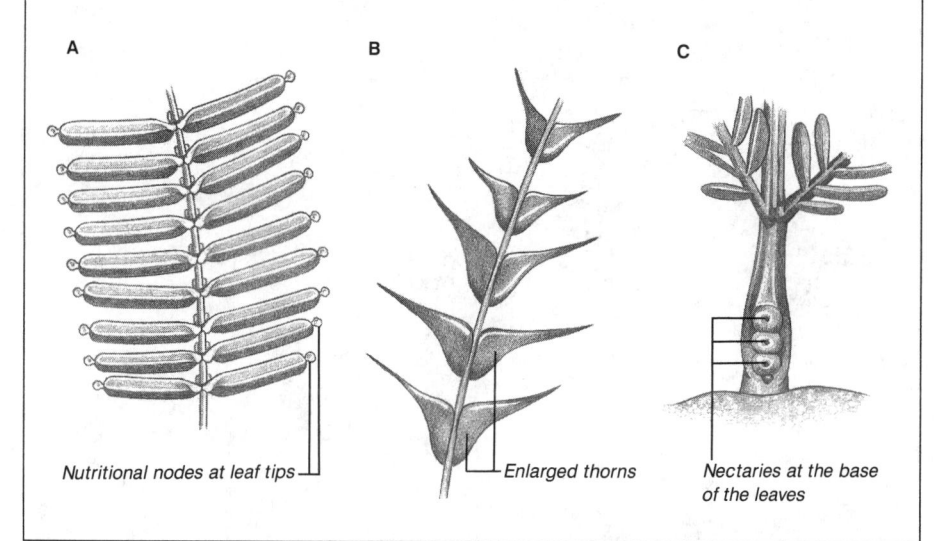

A — Nutritional nodes at leaf tips
B — Enlarged thorns
C — Nectaries at the base of the leaves

    c. Other examples of obligate nonsymbiotic mutualism include the relationship between ants and many plant species; one of the most dramatic examples is the ant-acacia relationship (see *Ant-Acacia Mutualism*)
       (1) Central American ants live in swollen thorns of the acacia tree *(Acacia)* and derive shelter and food from the plants
       (2) In return for food and shelter, the ants protect the acacia from herbivores and, in some cases, remove competing vegetation from around the trees
       (3) Neither the ants nor the trees can survive in the absence of the other
       (4) Aside from these cooperative aspects of their life cycles, the ants and acacia trees live independent lives
    d. The term "nonsymbiotic" in this context is somewhat misleading; as the above examples suggest, the species involved are in direct contact with one another (symbiosis, by definition) albeit only for a portion of their life cycles
4. In *facultative mutualism,* the partners are not completely obligated to live in a relationship with the other species, but they most often are found in association with one another
    a. The benefits of facultative mutualism often are spread over many groups of organisms
    b. Common examples of facultative mutualism can be found in pollination and seed dispersal relationships
    c. Cross-pollination requires pollen transfer from the anthers of one plant to the stigma of another plant of the same species

## Coevolution Among Flowers and Pollinators

Bee-pollinated flowers (A) are constructed so that the bee must travel past the stamen and pistil to reach nectar. In hummingbird-pollinated flowers (B), the stamen and pistil often hang down so that they touch the bird's head as it drinks the nectar. Beetle-pollinated flowers (C) generally are very open and provide easy access to the pollen.

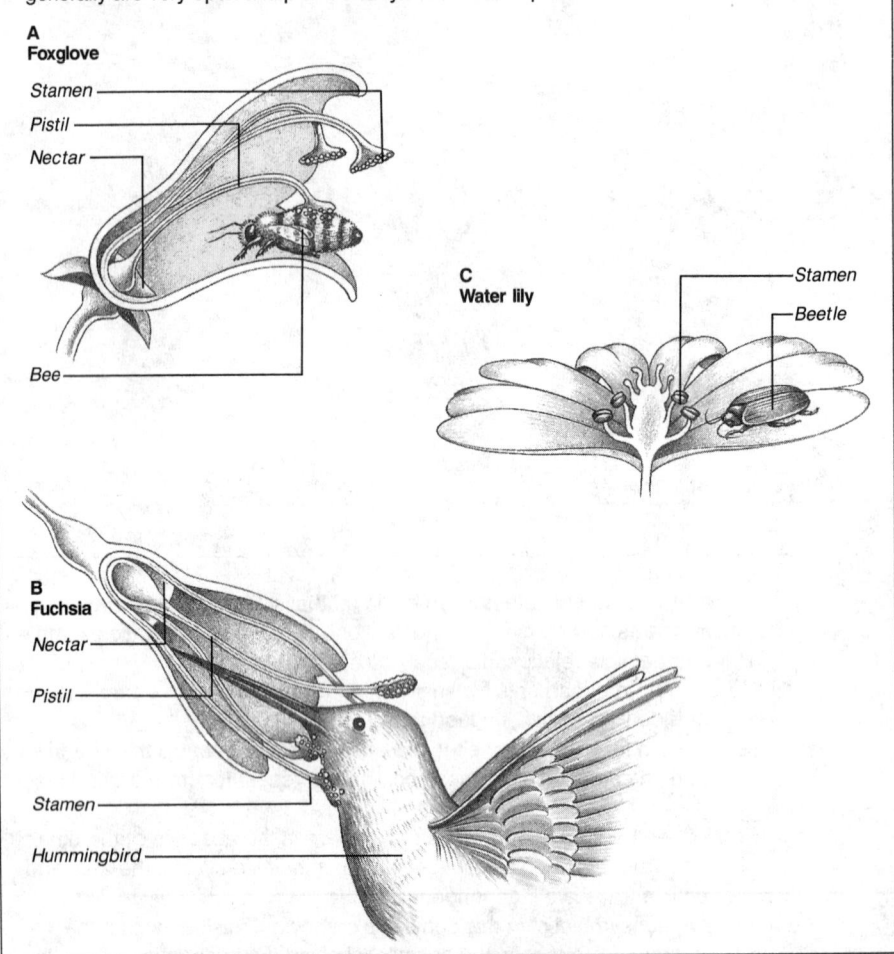

**A**
**Foxglove**
Stamen
Pistil
Nectar
Bee

**C**
**Water lily**
Stamen
Beetle

**B**
**Fuchsia**
Nectar
Pistil
Stamen
Hummingbird

(1) Wind dispersal is an effective means of pollen transfer when the plants grow in large, homogeneous stands, where the probability of wind-blown pollen landing on the pistil of another flower of the same species is high; however, such dispersal is ineffective when plants are scattered individually or in small patches across the landscape; these scattered plants must depend on animals (that is, insects, birds, or bats) for pollination (see *Coevolution Among Flowers and Pollinators*)
(2) A basic dichotomy exists with respect to pollination
    (a) A plant maximizes its fitness by attracting pollinators for frequent visits, which ensures effective pollen transfer

(b) However, donating nectar and pollen as rewards to the pollinators uses calories and nutrients that would otherwise be available to form seeds and fruits
- (3) From the perspective of a pollinator, the optimal strategy to enhance survival and fitness is to maximize the return in calories and nutrients with a minimal number of flower visits and minimal travel time between flowers
- (4) Some plants are generalists with respect to the pollinators they attract (for example, blackberry, elderberry, cherry, and goldenrod); these flower profusely, providing much nectar to attract a great variety of pollinators
- (5) Other plants are more specific or discriminating with respect to pollinator attraction and thereby ensure more efficient pollination; an example is the relationship between hummingbirds and the flowers they visit, such as the scarlet monkey flower or tropical varieties of the genus *Heliconia*

d. Plants with seeds too heavy to be dispersed by wind or water depend on animals to carry their seeds to new areas of establishment; these seed dispersal mechanisms therefore provide more examples of facultative mutualism between plants and animals
- (1) Aquatic plants may rely on birds for dispersal
    - (a) The bird consumes a fruit containing seeds, which are not affected by the bird's digestive tract
    - (b) The seeds pass through the bird and are deposited in another location away from the original plant
- (2) Fruits of flowering dogwood, spicebush, and wild grape ripen for consumption by migratory birds, which can scatter them widely
- (3) Small-seeded fruits, such as blackberries, blueberries, and mulberries, are sweet, fragrant, and appealing to small mammals and birds; these seeds pass quickly through the animals' gut
- (4) Larger-seeded fruits, such as cherries, are taken by such animals as raccoons, opossums, turkeys, robins, foxes, and skunks, which often distribute the seeds over wide geographic areas

## III. Other Species Interactions

### A. General information
1. The effects of interactions among populations can be positive, negative, or neutral in the impact on the interacting populations
2. Mutualism is a special type of relationship because of the coevolutionary aspects of the relationship and the positive benefits derived by both partners
3. Competition and predation are unique interactions in which one or both partner populations are harmed in the relationship; these relationships are explored in Chapters 7 and 8

### B. Commensalism
1. In *commensalism,* one population is favorably affected and the other is not affected

2. The one-sided relationship of commensalism is exemplified by epiphytic plants that grow in the branches of trees and depend on the trees only for support
3. The epiphyte benefits from the relationship by obtaining a substrate for growth; the tree is neither helped nor harmed

## C. Amensalism
1. In *amensalism,* one population is harmed and the other is unaffected
2. Amensalism typically involves a chemical inhibition by one of the organisms involved, often by the production of antibiotics or allelopathic chemicals that inhibit or interfere with the growth of a competitor
3. Most examples of amensalism can also be considered examples of interspecific competition

# Study Activities

1. Compare and contrast the terms coevolution, mutualism, symbiosis, facultative, and obligate.
2. Choose at least three examples of mutualism; explain the relationships involved and describe which category of mutualism is represented by your examples.
3. Briefly explain the origins and significance of mutualism.
4. Using the illustration in *Coevolution Among Flowers and Pollinators* (page 98) as a guide, explain or describe the coevolutionary adaptations among flowers and pollinators.
5. Explain commensalism and amensalism and give an example of each.

# 11

## Community Ecology Basics

### Objectives
After studying this chapter, the reader should be able to:
- Describe the importance and role of vertical stratification and horizontal heterogeneity in the structure of an ecosystem.
- Explain the ecological significance of ecotones.
- Describe the concepts of relative abundance, species dominance, diversity, and evenness.
- Explain the three typical distributions of organisms found in natural communities.
- Discuss the concept of biodiversity and explain its importance, the ways in which it is threatened, and how it might be preserved.

## I. Physical Structure of Communities

### A. General information
1. A *community* is a collection of several interacting populations located in a specific geographic area
2. Geologic formations (for example, rocks, sand, hills, mountainsides) and plants provide the physical strata on which organisms live and, as such, greatly determine the community's physical structure
3. A terrestrial community's structure and form primarily depend on its vegetation; for example, the structure of a grassland community is significantly different from that of a forest community because of the form of the predominant plant life
4. A complex physical environment with a variety of small habitats generally will have a wide variety of organisms, because the various habitats provide different places for organisms to live

### B. Vertical stratification
1. Plant life within a community often can be organized into layers or strata; this layering produces a variety of different areas (physical structure) in or on which organisms can live
2. In terrestrial communities, vertical stratification largely is determined by the size, branching, and types of leaves associated with the various plants of that community
3. A well-developed forest community has a highly stratified structure with many distinct layers

a. The forest vegetation consists of several layers, each of which provides habitat for animal life; these layers comprise the various plants that inhabit the forest
   b. The uppermost layer of vegetation, the *canopy,* consists of the upper leaves and branches of the tallest trees
      (1) The canopy may be fairly open, with many gaps through which sunlight can penetrate to the forest floor, or it may be closed, with dense layers of leaves that virtually block light from entering the community below
      (2) The degree to which the canopy is open or closed will have a significant impact on the type of plants and animals that inhabit the lower layers of the forest by altering such factors as sunlight availability, moisture conditions, and temperature
   c. Beneath the canopy, the *understory layer* consists of small, shorter trees
   d. Beneath the understory tree layer is the *shrub layer* consisting of woody vegetation tending to be less than 10 feet high
   e. The *herbaceous layer* is found beneath the shrub layer and consists of small, soft-stemmed plants such as wildflowers and ferns, as well as grasses
   f. The *forest floor,* the lowest layer of the forest vegetation, consists of fallen leaves, twigs, and branches, as well as small plants such as mosses and lichens
4. The stratification of aquatic communities is based largely on the depth of light penetration, temperature, and oxygen concentration
5. In general, the greater the number of distinctive strata in a community, the more diverse the animal life

## C. Horizontal heterogeneity
1. In terrestrial communities, vegetation is found in a mosaic or patchwork arrangement across the landscape; this pattern is referred to as **horizontal heterogeneity**
2. Horizontal heterogeneity results from the interaction of various environmental and biological factors
   a. Environmental factors that influence horizontal heterogeneity include soil structure and fertility, moisture conditions, light patterns, and runoff patterns
   b. Biological factors that influence horizontal heterogeneity include grazing by animals and competition among plant species
3. The influence of biological and physical factors on heterogeneity can influence recruitment, growth, and reproduction of plant species, which in turn influence animal populations

## D. Edges and ecotones
1. *Ecotones* are areas of transition between two or more ecosystems that form a defined edge, an gradual progression from one ecosystem to the other, or a mosaic of small patches between the two ecosystems; they arise from a blending of vegetational types
2. Ecotones can exist naturally or can be created by agricultural or landscaping practices
3. Ecotones are classified as inherent or induced depending on how they are formed and how long they persist

a. *Inherent ecotones* result from discontinuities in environmental features, such as abrupt changes in soil types, differences in drainage patterns, or other geographic or geologic factors
      (1) Vegetation types in inherent ecotones are determined by long-term natural features
      (2) Inherent ecotones generally are stable and more or less permanent
   b. *Induced ecotones* are produced by temporary landscape features resulting from short-term environmental disturbances, such as floods, erosion, fire, timber harvesting, agriculture, and grazing; generally, they change or disappear with time and can be maintained only by periodic disturbance or activity
   c. The richness (variety of kinds) of species in an ecotone is a result of the combination of flora and fauna from the adjoining communities and the addition of species favored by the edge environment
   d. The greater the contrast in communities forming the edge or ecotone, the greater the species richness within the ecotone
   e. A forest edge ecotone (the edge between a forest and a grassland community or a forest and a shrub community) is the favored habitat of many songbirds and game species, including deer, rabbit, pheasant, grouse, and squirrel

## II. Biological Structure of Communities

### A. General information
   1. The abundance of, relative proportions of, and interactions among species have a pronounced impact on the organization of a community
   2. In some communities a single species or group of species modifies the environment in such a way as to control the community; such controlling species are called dominants

### B. Species dominance
   1. **Dominant species** are those species or groups of species that biologically control a community by altering the environment
   2. A dominant species may be the most numerous, comprise the largest biomass, preempt the most space, make the largest contribution to energy flow, or exert great control over the movement of minerals and nutrients through the ecosystem
   3. In most communities, the dominant species are those whose removal would have a profound impact on the biotic community and microclimate and often are the species with the highest productivity in a given trophic level
   4. A species becomes dominant because it can effectively exploit a range of environmental conditions better than associated species
   5. To compare communities, ecologists calculate a measure of dominance using various mathematical formulas; one of the most common is Simpson's index of concentration:

   $$C = \sum_{i=1}^{N} (n_i / N)^2$$

   where C equals dominance or concentration value (Simpson's index), $n_i$ equals

## Calculation of Dominance and Diversity Measures

The chart below shows the calculation of dominance and diversity for two hypothetical communities, each of 10 species with a total of 100 individuals. The communities differ in the distribution of those individuals among the species. Simpson's index is a measure of dominance. As seen, Community 1 has a higher dominance value than Community 2. Diversity is measured by the Shannon index, which combines the two aspects of diversity, evenness or equitability and richness. Note that in this example, Community 2 is considered to display higher diversity. The species richness or number of species of both communities is equal. For the equations used to calculate the formulas below, see chapter text.

|  | NUMBER OF INDIVIDUALS | |
|---|---|---|
| Species | Community 1 | Community 2 |
| A | 91 | 10 |
| B | 1 | 10 |
| C | 1 | 10 |
| D | 1 | 10 |
| E | 1 | 10 |
| F | 1 | 10 |
| G | 1 | 10 |
| H | 1 | 10 |
| I | 1 | 10 |
| J | 1 | 10 |
| Total Individuals | 100 | 100 |

|  | Community 1 | Community 2 |
|---|---|---|
| Shannon's index (H) | 0.50 | 2.3 |
| Simpson's index (C) | 0.83 | 0.1 |
| Species richness (s) | 10 | 10 |

importance value for the i th species (as measured by biomass, productivity, or census), and N equals total of all importance values for the community
   a. Simpson's index ranges in value from 0 to 1
   b. As the index approaches 1, the community is considered to be dominated by the most important species (see *Calculation of Dominance and Diversity Measures*)

C. Species diversity
   1. The concept of species diversity has two components: *richness* (or species density) and **evenness** (a measure of relative importance)
      a. Species richness typically is measured as the total number of species present in a community
      b. Species evenness is a measure of the apportionment or allotment of individuals among species, based on relative abundance of species and the degree to which a species dominates the community
   2. One of the most widely used indices of diversity is the Shannon index, first proposed by mathematicians C.E. Shannon and W. Weaver in 1949

a. The Shannon index of diversity includes both species richness and equitability
b. The formula for the Shannon index is:
$$H = -\sum_{i=1}^{s} (p_i)(\log p_i)$$
where H equals diversity index (Shannon index), s equals number of species, i equals species number, and $p_i$ equals proportion of individuals of the total sample belonging to the ith species
3. Diversity indices may be used to compare species diversity within communities, between communities, and among communities over a wide geographic area
4. Local and regional diversity reflect different scales of measurement (smaller local scale vs. larger regional scale) and different environmental influences

## D. Species diversity hypotheses
1. Ecologists have proposed numerous hypotheses to explain why one area or region should hold more species than another, particularly the observation that tropical areas hold more species than temperate areas
2. The *evolutionary time hypothesis* suggests that diversity relates to the age of the community
   a. Evolutionarily older communities hold a greater diversity than do younger communities
   b. According to this hypothesis, tropical communities generally are older than temperate or arctic communities because the environment is more constant and climatic catastrophes are less likely, thus giving species a greater opportunity to establish and evolve
3. The *ecological time theory* proposes that time is required for species to disperse into areas of unoccupied habitat; as a consequence of this slow colonization process, temperate areas hold fewer species than tropical areas because temperate communities are likely to be evolutionarily younger than communities in tropical areas
4. The *spatial heterogeneity theory* suggests that more diverse and varied physical environments will support more diverse and complex plant and animal communities
   a. The greater the variation in topographic relief, the more complex the vertical stratification of vegetation and the more types of habitats available to organisms
   b. The more complex the physical environment, the more species an area will hold
5. The *climatic stability theory* proposes that more stable environments will hold more species
   a. The climatic stability theory is similar to the evolutionary time theory
   b. Under predictable and stable environmental conditions, species evolve narrow niches and specialized feeding habits, which often results in a great number of species in a particular habitat
6. The *productivity theory* suggests that the level of diversity in a community is determined by the amount of energy flowing through the food web; the more food produced (greater productivity), the greater the diversity
   a. Although this concept is true in a general sense, there are many exceptions

b. One notable exception to the productivity theory can be observed when an aquatic system experiences enrichment from sewage pollution; the enrichment increases productivity, but species diversity declines rather than increases
7. The *stability-time theory* rests on two contrasting types of communities: physically controlled and biologically controlled
    a. In physically controlled communities, organisms experience physiologic stress caused by fluctuating physical conditions
    b. In biologically controlled communities, physical conditions are relatively uniform and not critical in controlling species, but biological interactions, such as competition and predation, become important influences
    c. In physically controlled communities, diversity is kept low because of the impact of physiological stress on survival and reproduction
    d. In biologically controlled communities, diversity tends to be greater because the environment often is ameliorated by the presence of plants and animals creating conditions that enable some more sensitive species to survive; also, diversity tends to be increased by the specialization that usually results from competitive interactions in these biologically controlled communities
    e. No natural community is completely physically or biologically controlled; rather, natural communities occur along a spectrum and are influenced by the interaction of these two types of communities, which are theoretical extremes
8. Although these theories and hypotheses are useful in discussing the rise of diversity in various habitats and may provide insights into how communities develop, several critical weaknesses limit their applicability
    a. All of these hypotheses assume that the communities are at competitive equilibrium and therefore that change among competitors is zero; such a state of equilibrium is unlikely to occur in natural systems, which are evolving
    b. Many fluctuations in both the physical and biological components of communities will prevent them from reaching an equilibrium
    c. All of these hypotheses are difficult to test experimentally and, therefore, are of limited use

# E. Species abundance
1. **Species abundance** refers to commonness or the number of species and the number of individuals of each species found in a community
2. Species abundance is closely related to diversity, dominance, and density
3. Most often investigators are concerned with **relative abundance,** the proportion that a particular species contributes to the total abundance of a community
4. Distribution of relative abundance exists in three types: log-normal, broken stick, and ecological dominance
    a. In a *log-normal distribution,* a few species are very common in a community and many species are very rare
        (1) The log-normal distribution occurs when the numbers of individuals assigned to each species is tabulated, the range of numbers of individuals is divided into classes, and the upper limit of a given class is double the upper limit of the next lower class

      (2) The smallest class to which species can be assigned is the class that represents species of one or two individuals; the next larger class includes species with two to four individuals, then four to eight individuals, and so on, until the last class is defined as that class which includes the most abundant species present in the community
      (3) The borderline species are divided equally among the two bordering classes
      (4) Log-normal distributions occur in many bird, insect, and algae populations
      (5) Log-normal distributions are the product of random processes
         (a) They result when the abundance of each species population is determined at random
         (b) They demonstrate that the relative distribution and abundance of plants and animals typically are determined by random processes
  b. The *broken stick distribution* predicts small-scale patterns of relative abundance determined by species interactions, rather than random processes as in the log-normal distribution
      (1) Ecologist R.H. MacArthur first described the broken stick distribution in 1957 as a mechanism through which groups of species divide available resources among themselves
      (2) The resources shared among species of the community are considered a "stick" that is broken and divided among the species
      (3) MacArthur considered only one critical resource or niche dimension that the species shared
      (4) Broken stick distributions of abundance can be demonstrated for closely related species with synchronized life cycles that live together in small areas
      (5) Broken stick distributions occur in many populations of birds, mollusks, and small crustaceans
      (6) The broken stick distribution implies that some relationship other than randomness (or pure chance) exists among the species
  c. *Ecological dominance distribution* occurs when one or several species preempt community resources
      (1) An ecological dominance distribution results in a ranking of species from most important to least important
      (2) This distribution is based on the assumption that relative abundance is proportional to resource use by a population and implies competition among the species for particular resources
      (3) The ecological dominance distribution results as species are added to a community and take resources that are no longer available to other species that may be added to the community
      (4) Ecological dominance is demonstrated in temperate vegetation communities, in algae in lake communities, and in intertidal marine communities

# III. Biodiversity

## A. General information
1. **Biodiversity** refers to all the variety and variability in nature; it is concerned with the variety of organisms and the genetic diversity within each species
2. Most biologists agree that we do not currently have a complete picture of all the numbers and kinds of species that inhabit the earth; between 1.5 and 1.6 million species of organisms currently are known, but most authorities believe the total number of species may range between 30 and 50 million
3. An ever-increasing human population, with its expanding need for food, water, energy, space, and other resources, threatens the survival of other species that cohabit the planet
4. No precise estimate of species loss can be made, but the current rate of extinction of birds and mammals may be as much as 100 to 1,000 times greater than it was in recent centuries
   a. **Threatened species** still are relatively abundant in some parts of their ranges but have declined significantly in some areas
   b. **Endangered species** are those organisms in immediate danger of extinction because of extremely low populations
      (1) Conservationists estimate that at least 105 of the more than 250,000 flowering plant species are endangered
      (2) Terrestrial vertebrates (that is, mammals, birds, and reptiles) are experiencing an estimated extinction rate of 1 to 10 species per year
      (3) Aquatic vertebrates (that is, fish, amphibians, and some reptiles) probably experience an extinction rate of 10 to 100 species per year
      (4) The extinction rate for plants also is probably in the range of 10 to 100 species per year
      (5) Aquatic invertebrates (invertebrates of major concern include insects, crustaceans, and mollusks) may experience an extinction rate of 100 to 1,000 species per year; terrestrial invertebrates have the highest extinction rate — 1,000 to 10,000 per year

## B. Importance of biodiversity
1. Biodiversity is a source of genetic diversity
   a. Each organism is a unique combination of genetic characteristics that enable it to adjust to various environmental conditions
   b. All the genes of all the individuals in a population collectively form the **gene pool**
   c. Genetic variety in an organism's gene pool is vital if the species is to adapt to changing environmental conditions
   d. The gene pool is the "raw material" of evolution and natural selection and, as such, provides the raw material for change and adaptation of a species
   e. Scientists believe that the gene pools of the earth's many species include undiscovered traits that may be important for the future of humans
   f. The potential importance of this natural biodiversity (and its accompanying gene pool) can be seen in the observation that approximately 45% of all medical prescriptions written in the United States contain at least one product of natural origin
      (1) Digitalis, used to treat chronic heart aliments, is obtained from the foxglove plant

- (2) Morphine, used to reduce pain, is extracted from the poppy plant
- (3) Taxol, which appears to be useful in treating certain forms of cancer (for example, ovarian cancer), is isolated from the bark of the Pacific yew tree
- (4) Current estimates suggest that only about 5,000 of the world's 250,000 species of flowering plants have been screened for possible medical applications
- g. Maintenance of large gene pools is valuable for agriculture
  - (1) Virtually all agricultural crops and livestock currently used by humans are domesticated native plants and animals
  - (2) Agricultural scientists need and use wild populations of native species to provide genetic characteristics to solve present and future food production problems (for example, pest resistance and immunity to pathogens)
2. Biodiversity is important in preserving the integrity of ecosystems and the functions they perform
   - a. Species participate in the various ecological processes that take place within and between ecosystems
   - b. Each species contributes to one or more ecosystem functions — production, decomposition, nutrient cycling, soil production, erosion control, pest control, water movement, and climate regulation
   - c. The many functions performed by the various species within the many ecosystems of the earth are important to the survival of an ecosystem, and collectively they are vital to the overall health of the biosphere and human survival
   - d. Some species play particularly vital or key roles in their ecosystems and consequently must be protected and preserved to maintain the ecosystem
     - (1) The American alligator is an example of a key species that plays a vital role in controlling the relative abundance of other species in the Everglades ecosystem; elimination of the alligator could have resulted in dramatic changes in the size of other populations within the Everglades and possibly led to wholesale disruption of the entire ecosystem
     - (2) As another example, the sea otter plays a key role in determining the presence or absence of other sea animals along the Pacific coast of North America
3. Biodiversity can have significant economic value
   - a. Humans use a vast variety of natural products in industry and agriculture, worth trillions of dollars in the world economy
   - b. Tourism and recreation in natural areas, parks, and preserves also have significant economic value
   - c. Sport hunters and fishers spend hundreds of millions of dollars annually
4. Biodiversity is important for aesthetic value
   - a. Many people derive a sense of pleasure or renewal from wilderness experiences or from viewing natural wonders, wild animals, and plants
   - b. Although some are moved to preserve biodiversity on aesthetic grounds, most humans — particularly those of poorer nations, the unemployed, or the hungry — are not likely to view nature's aesthetic properties as valuable
5. Biodiversity has intrinsic value

  a. Many authorities agree with David Ehrenfeld, who argues that by assigning economic, gene pool, or ecosystem function values to biodiversity, we legitimize its very destruction because of human socioeconomic behaviors that emphasize short-term profit and net return
  b. Short-term economic gain from destruction of biodiversity is the driving force behind the extinction of many species
  c. According to Ehrenfeld and others, the value of biodiversity cannot and should not be couched in terms of human values; biodiversity does not exist for humans to approve or disapprove
  d. Based on this line of reasoning, biodiversity should be maintained because it is there, not because humans value it or may profit from it

## C. Conservation and preservation
1. To evaluate possible methods of conservation and preservation, it is important to examine the ways in which humans influence the extinction of species
  a. Wildlife loss caused by human activity can occur in three ways: habitat destruction, extinction as a result of direct exploitation of plants and animals, and the introduction of exotic species into new areas
  b. Until a few centuries ago, the impact of human populations on both the environment and cohabiting species was minimal because of our small numbers, our hunter-gatherer society, and wide global population dispersal
  c. In modern times, destruction of natural habitats caused by human population and economic pressures is a major threat to biodiversity, probably accounting for 67% of the species now listed as endangered by the International Union for Conservation of Nature and Natural Resources
  d. Destruction of an area, such as a tropical rain forest or natural prairie, not only destroys plant life but also (by eliminating or drastically altering the habitat) often destroys animal populations
  e. Today, approximately 50% of the tropical forests have been destroyed; in the United States alone, more than 60,000 acres of ancient trees are cut each year in the Pacific Northwest
  f. Some scientists warn that even the maintenance of preserves and natural areas may not be sufficient to prevent dramatic loss in biodiversity, because the preserves themselves may not be large enough, of sufficient quality, or of the appropriate ecosystem type
  g. Commercial hunting and fishing are examples of direct exploitation of species that poses a significant threat to biodiversity
    (1) In the past, a significant number of species became extinct or endangered because of their importance as food or because their killing was considered sport; examples include the passenger pigeon, great auk, heath hen, wild turkey, and sandhill crane
    (2) Up to 30% of all international trade in plants and wildlife may be illegal, with products used for skins, fur, pets, and jewelry
  h. The accidental or intentional introduction of alien or exotic species has caused some serious problems
    (1) When an alien species is introduced into a new environment, the alien may be able to flourish because of the lack of the competition and predation that controlled its numbers in its resident ecosystem
    (2) If the alien species is a superior competitor or predator, it may eliminate vast numbers of species from the new ecosystem

2. Efforts to preserve biodiversity are difficult and will require much effort, time, and funds and various approaches
   a. Efforts to maintain biodiversity have focused on individual species, through efforts to prevent further extinction or to reestablish lost populations
      (1) The U.S. Fish and Wildlife Service has approved more than 310 recovery plans of action for helping species listed as endangered or threatened; however, more than 3,600 species await consideration
      (2) Species-level approaches include seed and gamete storage, maintaining plants and animals in zoos or conservation parks, cryogenic freezing of plant and animal tissues, and captive breeding and release programs
      (3) Many biologists feel that the species-level approach to preservation does not adequately consider the vital ecological relationships important to species or ecosystem survival
      (4) Because of their high cost (that is, in time, money, and other resources), species-level approaches appear to inevitably require choices of which species to save and which to abandon and thus are considered by some to be undesirable or stop-gap at best
   b. A more holistic approach to preserving biodiversity is through a landscape or habitat preservation and restoration mode
      (1) In biological preserves, plants and animals have sufficient space and resources to survive and maintain their populations
      (2) In preserving entire ecosystems (rather than attempting to save individual species), the connections and interactions between species are preserved, and the possibility of survival of all species is enhanced
      (3) The hope is that by maintaining sufficiently large tracts of intact ecosystems carefully chosen for their biological and physical characteristics, the ecosystems can self-perpetuate and survive
      (4) Much of the controversy over landscape or preserve-based conservation arises when human needs conflict with the need to preserve a particular habitat or species
      (5) A major problem with biological preserves is that scientists are not sure how large they should be and how many are needed to adequately preserve biodiversity
      (6) Biological preserves also need to be connected to one another, to allow migration from one area to another; these corridors between preserves maintain genetic diversity and healthy populations

## Study Activities

1. Sketch and label the various layers of vertical stratification found in a temperate forest ecosystem.
2. List and describe the common measures of species dominance and diversity; give the mathematical formulas used and describe what these formulas reveal.
3. Give an example of an ecotone. Explain why such habitats are important.
4. Compare and contrast the three distributions of organisms typically found in natural systems: log-normal, broken stick, and ecological dominants.
5. List two reasons why humans should preserve biodiversity and justify them.

# 12

# Succession

## Objectives

After studying this chapter, the reader should be able to:
- Define and give examples of primary and secondary succession as it occurs in various ecosystems.
- Describe the three models of succession as proposed by Connell and Slatyer.
- Explain the overall changes in ecosystem attributes that occur as an ecosystem passes through the various stages of succession.
- Explain the historical perspective of Clements and Gleason in the development of the concept of succession.
- Describe the concept of island biogeography and its impact on an area's species composition.

## I. Processes of Succession

### A. General information
1. **Succession** refers to the gradual change occurring in an ecosystem of a given area on which populations succeed each other
2. Succession generally is a plant-driven or plant-dominated process in which animal communities change as the plant community changes
3. Succession occurs in two general forms: primary and secondary
    a. **Primary succession** usually involves colonization of bare ground where no previous ecosystem was present; examples include sand dunes, volcanic flows, mud flats, glacial till, and dystrophic lakes and ponds
    b. **Secondary succession** is characterized by replacement of a community after some disturbance where a previous community existed on the site; examples include old abandoned fields, wind-blown gaps in forests, and areas after fire
4. Succession also may be categorized based on the origin of factors that control the successional sequence
    a. Succession is considered *autogenic* when directed from within the ecosystem, where changes in habitat are brought about by actions of the biota
        (1) Autogenic succession occurs as soils accumulate and nutrients collect
        (2) Changes in the habitat result in community replacement
        (3) Many terrestrial ecosystems are thought to be largely autogenic
    b. Succession is considered *allogenic* when driven by forces or factors external to the ecosystem

(1) Aquatic systems are thought to be largely allogenic
(2) Aquatic plants and animals respond to changes from outside, as in the case of pollution inputs or progressive drop in water table in a marshland as a result of draining

## B. Stages of succession
1. Ecologists describe the successional sequence of an area in terms of the major stages of characteristic vegetation
2. The earliest stage of succession, which contains the first colonizers of an area, is known as the *pioneer stage*
3. Each subsequent stage of succession, called a *sere,* is a point in the continuum of vegetation through time and is recognizable as a more or less distinct community with its own characteristic structure and species composition
4. Seral stages may last for brief periods or may persist for many years; in some cases, seral stages may be missed completely
5. Eventually, succession slows and the plant community achieves some degree of equilibrium or steady state with the environment in which the plant and animal communities appear to persist; this mature, relatively self-sustaining seral stage is known as the *climax stage* or *climax community*
   a. Traditionally, the climax community has been considered an endpoint of succession and thought to be characteristic for a particular set of environmental or climatic factors in a particular geographical area
   b. More recently, ecologists tend to deemphasize the finality of the climax stage and consider it in terms of an evolving and changing community that may be less permanent or persistent than previously thought
6. There are three theoretical approaches to the concept of the climax: monoclimax theory, polyclimax theory, and climax pattern theory
   a. The *monoclimax theory* recognizes only one climax in a given area, whose characteristics are determined by climate
      (1) Successional processes are thought to modify the environment and overcome the effects of differences in topography and the soil, resulting in a very predictable endpoint community
      (2) According to the monoclimax theory, if given sufficient time all seral communities will progress to the climax community characteristic of that area
   b. The *polyclimax theory* suggests that the climax community of a region comprises a mosaic of different vegetation types controlled by soil nutrients, soil moisture, topography, slope, exposure, fire, and animal activity
   c. The *climax pattern theory* states that an ecosystem's total environment determines the composition of the climax community
      (1) The climax community arises from the interaction of biotic factors and abiotic factors in a particular region
         (a) Biotic factors include the availability of plants and animals to colonize an area and the chance arrival of seeds and animals
         (b) Abiotic or physical factors include soils and climate
      (2) Biotic and abiotic factors are likely to change with time and result in subsequent changes in the community
      (3) The climax is that community that is most widespread and prevailing at a given time

7. In some cases, a successional sequence is held or arrested at one stage and never proceeds to climax because of repeated disturbances, such as fires, grazing, or power line right of ways; such an arrested stage is called a ***proclimax***

## C. Terrestrial primary succession
1. On terrestrial sites where primary succession occurs, no soil exists initially
2. This type of succession generally occurs on rock outcrops, cliffs, volcanic deposits, newly deposited alluvial soils on floodplains, and glacial till
3. Glaciers at Glacier Bay, Alaska, have been monitored by investigators since 1890 and provide a good example of terrestrial primary succession
    a. As glaciers recede, they leave behind a gravel mixture that generally is devoid of organic material, contains no nitrogen, and is alkaline (pH ranging from 8.0 to 8.4)
    b. The empty habitat created by the retreat of the glaciers subsequently is colonized
        (1) Succession begins with a pioneer community of arctic herbs and dwarf willows, which later is replaced by willow shrubs; these, in turn, are replaced after 50 years by an almost pure stand of alder bushes
        (2) The pure stand of alders gradually is invaded and later replaced by Sitka spruce; after 120 years, a conifer forest develops on the site
        (3) The spruce forest gradually is invaded by hemlock for the next 80 years; the resulting climax community, a mixed spruce-hemlock forest, is established approximately 200 years after the glacial till first was deposited
    c. Profound changes in the physical habitat occur during this 200-year succession
        (1) During the successional sequence, soil is developed and amended by the accumulation and decay of leaves and other debris from the vegetation of each seral stage
        (2) The soil pH change (that is, significant pH drop) is most profound during the seral stage in which alders are dominant; this probably is caused by the deposition and decomposition of acid leaf residues by the alders
        (3) Organic matter accumulates during this succession, in large part because of the declining pH, although a more or less constant level of soil organic matter may be reached when the community reaches an age of 100 years
        (4) Soil nitrogen levels begin to rise during the pioneer stage, when some nitrogen fixation occurs; the largest increase in nitrogen occurs when alders predominate, because they have prominent root nodules containing nitrogen-fixing bacteria
        (5) Soil nitrogen levels begin to drop as the alders are replaced with spruce; this probably reflects the fact that spruce do not fix nitrogen
        (6) Studies indicate that the spruce-hemlock forest (climax forest) cannot grow on the glacial till as it is initially left by glaciers; rather, these species can colonize the area only after the soil pH decreases and the soil nitrogen level increases — factors that are altered by the successional sequence

4. Along the southern and eastern shores of Lake Michigan, sand is deposited by the lake waters and thrown up into dunes by wind action; once deposited, this sand becomes a habitat that experiences succession as a community develops
   a. Sand is deposited around some object (for example, a snow fence, a pile of trash, or a large rock) that presents an obstacle to the wind; the wind is slowed and the sand it is carrying is dropped
   b. Two species of grass — beach (marram) grass and dune grass — generally are the pioneer species first established on a newly formed dune; both species are perennial plants and can produce new roots from buried stems and send up new shoots above the sand
   c. Marram and dune grasses quickly are invaded by sand cherry and cottonwood, shrubs that can survive on the accumulating and shifting sand
   d. The dune is stabilized by the growth of marram grass, dune grass, sand cherry, and cottonwood
   e. At this point, the dune may proceed through other seral stages and eventually develop into a forest; it also may become a "wandering dune" if the vegetation cannot stabilize the sand, which then is moved by wind erosion
      (1) The cause of the wandering dune phenomenon is not well understood, but destruction of the existing vegetation apparently is important
      (2) Most commonly, storm damage destroys vegetation and wind blows away the exposed sand, undermining the vegetation and removing even more sand
      (3) The sand may be blown back across the landscape and may cover other communities that have developed
   f. If marram and dune grasses, along with sand cherry and cottonwood, establish sufficiently to stabilize the dune, then further colonization by shrubs such as bearberry, juniper, dogwood, cherry, and willow occurs
   g. At this point, cottonwood trees may predominate, and a cottonwood stage may develop
   h. If a cottonwood stage does not develop, then the shrub stage often is invaded by jack pines and occasionally other pines
   i. Eventually, the cottonwood or jack pine stage is invaded by oaks, which germinate and grow among the other trees; the climax community then is an oak forest
      (1) At the southern end of Lake Michigan a low, open black oak forest is the climax community
      (2) Along the northern and eastern shores, red oaks have invaded the cottonwood or jack pine stages and eventually have been invaded by basswood, sugar maple, beech, and hemlock; the climax community contains all five species
   j. If this forest were disturbed (for example, from storm, fire, or human activity), then the successional sequence would begin again with marram and dune grasses
   k. Within this basic successional framework, the communities of the sand dunes form a mosaic of embryonic dunes, wandering dunes, and stabilized dunes, all in various stages of succession from marram grass to forest
   l. This basic successional pattern also is visible moving inward away from the lake shore, with early seral stages nearer the shoreline and older climax communities further inland

## Secondary Succession in Temperate Terrestrial Ecosystems

Secondary succession in temperate areas is illustrated below. The successional sequence begins with a disturbance that removes vegetation and ends with a mature deciduous forest.

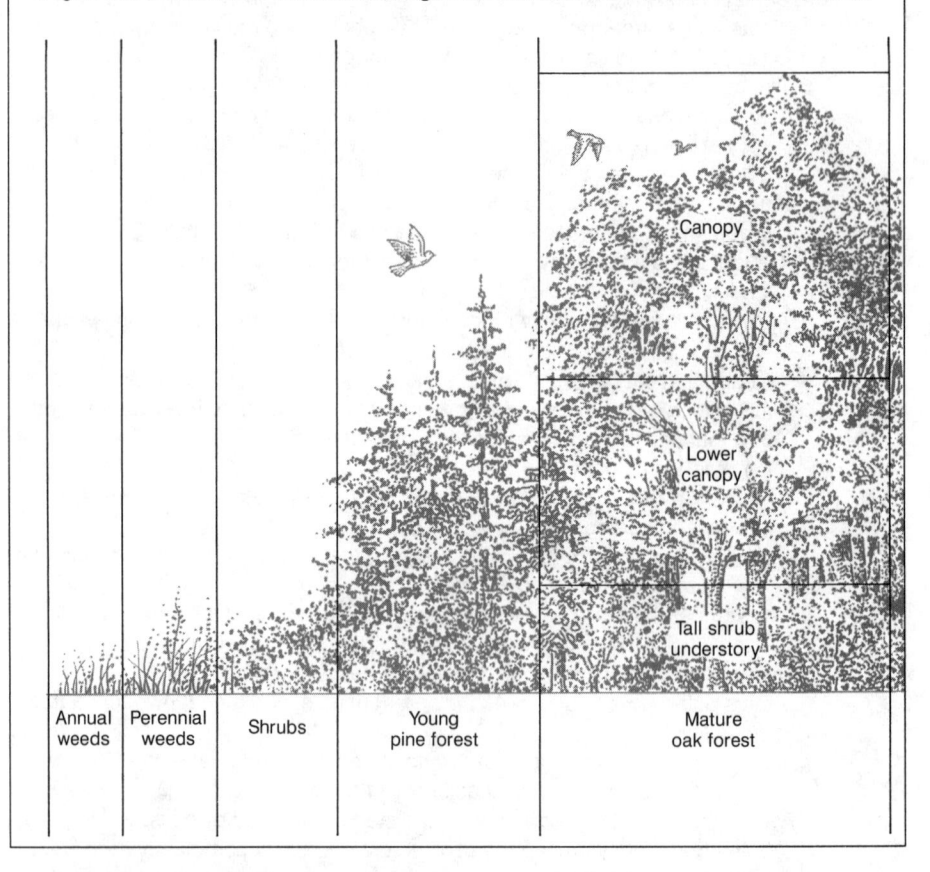

| Annual weeds | Perennial weeds | Shrubs | Young pine forest | Mature oak forest |
|---|---|---|---|---|

Mature oak forest layers: Canopy; Lower canopy; Tall shrub understory

**D. Terrestrial secondary succession**
1. Secondary succession most commonly occurs on abandoned land, waste areas, and areas disturbed by human activity or by natural events, such as fire
2. The plant species most likely to colonize an area during secondary succession are weeds
3. Typically, in secondary succession the seeds of colonizers remain viable for long periods and may remain in the soil for an extended time
4. Rapid and successful colonization often is accomplished by efficient seed dispersal
5. A typical example of secondary succession is old field succession (succession on abandoned agricultural land) in temperate areas of the southeastern United States (see *Secondary Succession in Temperate Terrestrial Ecosystems*)
    a. In the first year that a cultivated field is left fallow, the site generally is colonized in the spring by annual species, such as crabgrass; by late summer, horseweed seed invades the field and grows to the rosette stage by late autumn

b. In the spring of the second year, horseweed dominates the field; during the summer other plants, such as asters and ragweed, invade the site
   c. The self-inhibitory effects of horseweed and the competitive superiority of asters result in the asters gaining dominance
   d. By the third summer, perennial and biennial plants begin to move in; these include broom sedge, goldenrod, Queen Anne's lace, wild strawberry, and bedstraw
   e. Also during this third year, pine tree seedlings begin to establish; within 5 to 10 years, the pines are tall enough to shade some of the perennial herbaceous species
   f. A layer of partially decomposed pine needles prevents pine seeds from reaching the mineral soil to germinate; coupled with competition for moisture and light from parent trees and seedlings of deciduous trees, this inhibits pines from regenerating themselves
   g. Deciduous hardwood trees grow up through the pine canopy and take over the site
   h. The development of a hardwood forest continues as shade-tolerant trees and shrubs fill the understory; at this point (approximately 100 to 110 years since the field was first abandoned) the seral sequence reaches the climax stage, in which only the dominant species of the upper canopy can replace themselves in their own shade
  6. Although the successional sequence just described may occur to some extent in secondary succession in temperate areas, many variations are possible, and the rate of development and particular species involved in each seral stage are influenced by local climatic, geologic, and geographic characteristics

## E. Aquatic succession
  1. The transition from pond to terrestrial community is an example of primary succession (see *Aquatic Succession,* page 118)
  2. Succession starts with open water and few, if any, living organisms
   a. The pioneer species generally are **plankton,** small floating or weakly swimming plants and animals, which may grow so densely as to cloud the water or form large floating mats
   b. Plankton may enter a water body attached to animals (feet of birds, hair of beaver) or attached to leaves or twigs that may be carried into the water body by streams or runoff from surrounding areas
   c. As plankton growth increases, other organisms (such as caddis flies, bluegills, sunfish, and bass) may colonize the pond by entering through a tributary stream
   d. Because ponds exist in low-lying areas, they serve as catch basins and settling areas for deposition of sediment from the surrounding landscape
   e. The collection of sediment forms a muddy **substrate** for rooted algae and aquatic plants
   f. The rooted plants, in turn, bind the loose sediments together and add organic matter
   g. The accumulation of layer upon layer of sediment and organic matter begins to reduce water depth, and the pond basin fills in
   h. Emergent and submerged vegetation continues to grow and further enrich the water with organic material

## Aquatic Succession

Lakes and ponds are located in low-lying areas and receive drainage and runoff from the surrounding landscape. Soil accumulates, terrestrial plants invade, and a mature forest eventually develops.

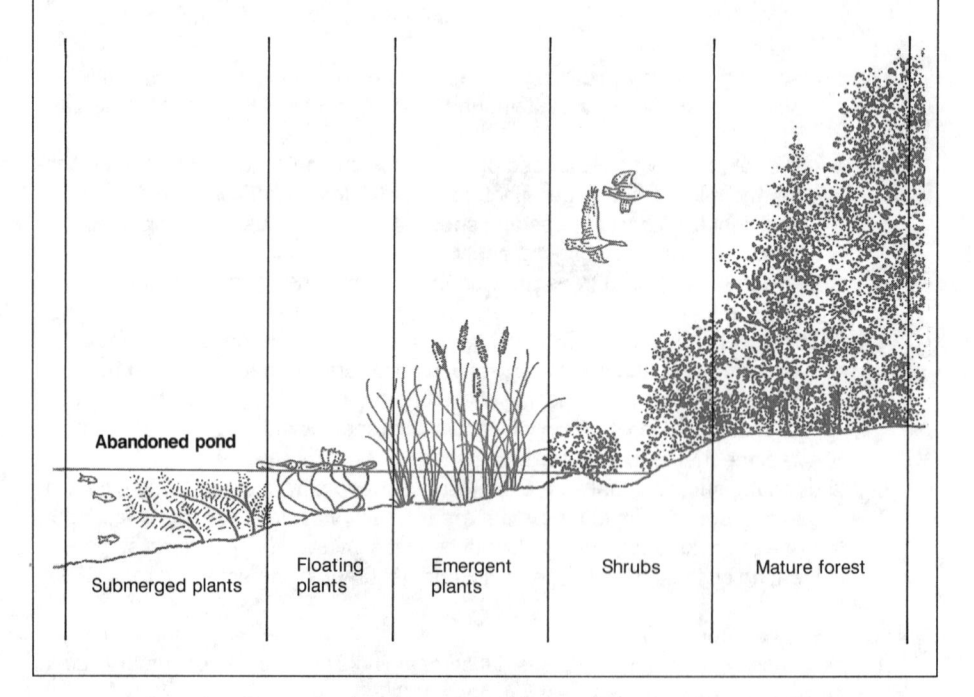

**Abandoned pond**

Submerged plants | Floating plants | Emergent plants | Shrubs | Mature forest

       i. As the water depth decreases, the pond supports emergent vegetation, such as cattails and sedges, and becomes a marsh
       j. As the substrate continues to be elevated by the accumulation of material produced or trapped by vegetation, land builds and the soil rises above the water table
       k. As soil is exposed to the air and dries, meadow grasses begin to invade the marsh
       l. Depending on local climatic and geologic conditions, the area will develop into either a grassland, a bog, or a forest
   3. Aquatic succession largely is self-driven by the changes induced by the vegetation and thus is considered largely autogenic

## II. Models of Succession

### A. General information
   1. Ecologists have attempted to develop models that explain the mechanisms by which succession occurs
   2. Historically, the observation of plant distributions led to the formulation of two opposing views of the mechanisms of succession

a. Plant ecologist F.E. Clements (1916) is credited with first advancing succession as an ecological concept
    (1) Clements viewed succession as a replacement process in which different stages of plants alter the environment to prepare the way for incursion by other species
    (2) According to Clements, one species replaces another until a climax community is obtained
    (3) Clements considered a climax community a "superorganism" able to self-perpetuate and reproduce on a particular site and thus maintain itself; as a superorganism the community will self-regulate and maintain itself at an equilibrium
    (4) Clements felt that as an ecosystem develops during the successional sequence, predictable changes in ecosystem functioning could be identified (for example, increased nutrient cycling, increased species diversity, and accumulation of biomass; also, as the community matures, costs of respiration begin to equal costs of production)
    (5) According to Clements, a community is an assemblage of closely linked species more or less connected by biotic interactions that cause the community to function as an integrated unit
b. Botanist H. Gleason (1917) challenged Clements' views, describing the mechanisms of succession in terms of random collections of plants
    (1) According to Gleason, a community is the result of a random process in which short-lived colonizing species eventually are replaced by longer-lived species
    (2) The pioneer species are those that arrived first on a site and are able to establish themselves under existing environmental conditions
    (3) The final outcome, and thus the resulting community, is based on local environment and climatic conditions and the interactions among the randomly arriving species
    (4) In general, Gleason believed that successional changes could be explained by population dynamics, such as competition and survivorship; thus, succession is nothing more than a process of species interacting at different stages in the development of a community
    (5) Gleason basically viewed a community as a chance assemblage of species found in an area because they happen to have similar abiotic requirements
    (6) Gleason's view of succession did not include a final climax or equilibrium point; rather, he envisioned a changing and dynamic, nonequilibrium community
3. The views of Clements and Gleason are incorporated into the current views of succession, which generally acknowledge a combination of both extremes in most succession sequences

## B. Connell-Slatyer models of succession
1. In 1988, two eminent ecologists, J.H. Connell and R.O. Slatyer, proposed three different models of succession: facilitation, tolerance, and inhibition
    a. These models attempt to explain the mechanism by which communities change over time and how these changes come about
    b. The facilitation model is basically a Clementsian view of succession

c. The tolerance model and the inhibition model include more of the effects of competition and the role of colonization in determining species composition and, as such, are more Gleasonian in approach
2. The *facilitation model* of succession describes classical succession, which is autogenic and largely controlled by changes brought about from within the ecosystem by the organisms themselves
   a. Facilitation succession generally is primary succession
   b. As abiotic factors are changed by the resident species, their influence on the community changes and thus the community changes
   c. The ecosystem continues to change until an equilibrium is reached
   d. The pioneer community is subject to wide ranges of temperature and moisture conditions
   e. Pioneer species are generalists and are ephemeral, quickly giving way to plants of more advanced seral stages
   f. The productivity and biomass of the pioneer community are low, but the ratio of productivity to biomass (the P/B ratio) is high, suggesting that energy transfer is inefficient
   g. Diversity initially is very low and increases in subsequent seral stages
   h. The changes in habitat brought about by the pioneer species enable other species to enter the community
   i. The climax community possesses higher biomass, greater species numbers, lower P/B ratios, and better-developed food webs than the pioneer community
3. The *tolerance model* of succession involves the interaction of life history patterns of species, especially the impact of competition
   a. In this model, pioneer species are those that can survive in the environment as it is first presented; but any species may be a pioneer, even later-seral stage plants
   b. Later-stage organisms may be present early on but do not characterize the community
   c. In this model, the pioneer species do not alter the abiotic environment in a way to prepare it for other colonizers; actually, the environment is likely to become less hospitable to new colonizers of earlier successional varieties as the community matures and plants grow to form a dense layer or canopy
   d. Later-seral stage species are tolerant of the conditions produced by this early seral assemblage of organisms
   e. Juveniles of later successional stages either are added by recruitment or are already present; in either case, they continue to grow and develop so that in time the early species will be eliminated
   f. Old field succession is thought to proceed according to the tolerance succession model
      (1) In old field succession, the seeds of trees and grasses already are present on the site
      (2) Initially, the most prominent plants are grasses, wildflowers, and weeds, with tree seedlings often present but inconspicuous
      (3) Shade beneath the pioneer plants inhibits the reproduction of most other pioneers, but later-stage plants that can tolerate the shade will gradually become more prominent in the community

g. The trends in biomass, productivity, and P/B ratio are consistent with those of the facilitation model
4. The *inhibition model* of succession is based exclusively on competition
   a. In inhibition succession, early successional species inhibit further change of the community by preventing or interfering with the establishment of new species
   b. As individuals die, they are replaced by others of either the same species or another species of a later seral stage
   c. The rate of change in the community is determined by the life span of the species involved, because replacement can occur only on removal of an individual already present
   d. Allelopathic interactions may be involved as the mechanism by which an earlier colonizer inhibits the establishment or growth of later arrivals
   e. The proclimax communities of shrubs that often develop in power line rights of way, which may suppress tree growth for as long as 40 years, are good examples of inhibition succession

## III. Changes in Ecosystem Attributes

### A. General information
1. Regardless of the mechanism of succession or the degree to which succession is considered equilibrium-based (after Clements) or nonequilibrium-based (after Gleason), certain general trends appear to be nearly universal
2. The trends that generally occur with succession are properties of the ecosystem as a whole; they are not necessarily true of every species in the community nor must they be true of every successional sequence

### B. Trends in community attributes
1. As a community progresses through the various seral stages, community biomass and organic detritus increase
2. Gross production increases with the community's age in primary succession, but little change occurs in secondary succession
3. As a general rule, net production decreases as the community ages
4. The productivity to respiration (P/R) ratio approaches 1 as the community approaches the climax stage
5. As a community proceeds through the various seral stages, nutrient cycles (that is, nitrogen, sulfur, phosphorus) become increasingly closed, so that very little of these nutrients leak from the ecosystem; similarly, turnover time and storage of essential elements increase, as do nutrient retention and conservation
6. Species composition changes with each seral stage of succession
   a. Diversity tends to increase during succession, both in species richness (variety) and evenness (equitability)
   b. During succession, *r*-selected species tend to be replaced by *K*-selected species (see Chapter 9, Life History Patterns, for a discussion of *r*- and *K*-selection)
   c. Life cycles increase in length and complexity in later seral stages
   d. The size of organisms, their propagules, or both tend to increase during succession

7. The stability of each seral stage is different, and the overall stability of the entire ecosystem changes over time
   a. *Resilience,* the ability to return to equilibrium after a disturbance, tends to decrease with successional age
   b. *Resistance* or persistence, inertia or the ability to resist perturbation, increases with successional age

# IV. Island Biogeography

## A. General information
1. The theory of *island biogeography* first was formulated by ecologists R.H. MacArthur and E.O. Wilson in 1963 to explain the rate of colonization and establishment of species and the ultimate species composition on islands
2. According to this theory, the number of species on an island will reach an equilibrium when rates of immigration and extinction are balanced
3. Although island biogeography was devised to explain colonization and community composition on islands, it also can be applied to patches of habitat on mainland continents, such as mountaintops, ponds, dunes, nature reserves, and highly disturbed areas

## B. Theory of island biogeography
1. In general, the theory of island biogeography states that the number of species on islands is the result of a balance between immigration and extinction (see *Rate of Colonization and Extinction in Island Biogeography*)
2. On islands too small to support speciation (formation of new species through adaptation and evolution) through geographical isolation of populations, the number of species increases via immigration from other sources, such as other islands or the continents
3. Although the rate of immigration may stay the same, successful establishment of new arrivals of new species to an island from a pool of mainland colonists will decrease as the number of species on the island increases
   a. Establishment declines because the greater the "pool" (or mainland) species found on the island, the fewer immigrants will belong to new species
   b. Later immigrants may be unable to establish populations because available habitats have already been filled or available resources already used
4. When all the species from the potential pool of colonists are found on the island, the species establishment rate of new species will be zero
5. As the number of species on the island increases, the rate of extinction will increase, because each population encounters stronger biotic challenges, such as competition, predation, and disease
6. Because many factors are likely to influence immigration and extinction rates, these rates probably do not vary in strict proportion to the number of potential colonists and the number of species established
   a. Some species probably are better colonizers than others and may reach the island first, so that the rate of immigration to and successful establishment on the island initially decreases faster than it would if all of the potential colonizers had equal potential for dispersal to and arrival on the island
   b. Biotic factors, such as competition and disease, are likely to cause extinction curves to rise progressively faster as species diversity increases because

## Rate of Colonization and Extinction in Island Biogeography

Initially, all arrivals or immigrants (I) are "new" to the island and therefore colonization rates are high. As a greater number of species establish on the island, fewer arrivals will be "new" and so colonization rates drop. At the same time, extinction rates (E) begin to rise. The point at which the extinction and immigration rates are balanced determines the equilibrium number of species that will inhabit the island.

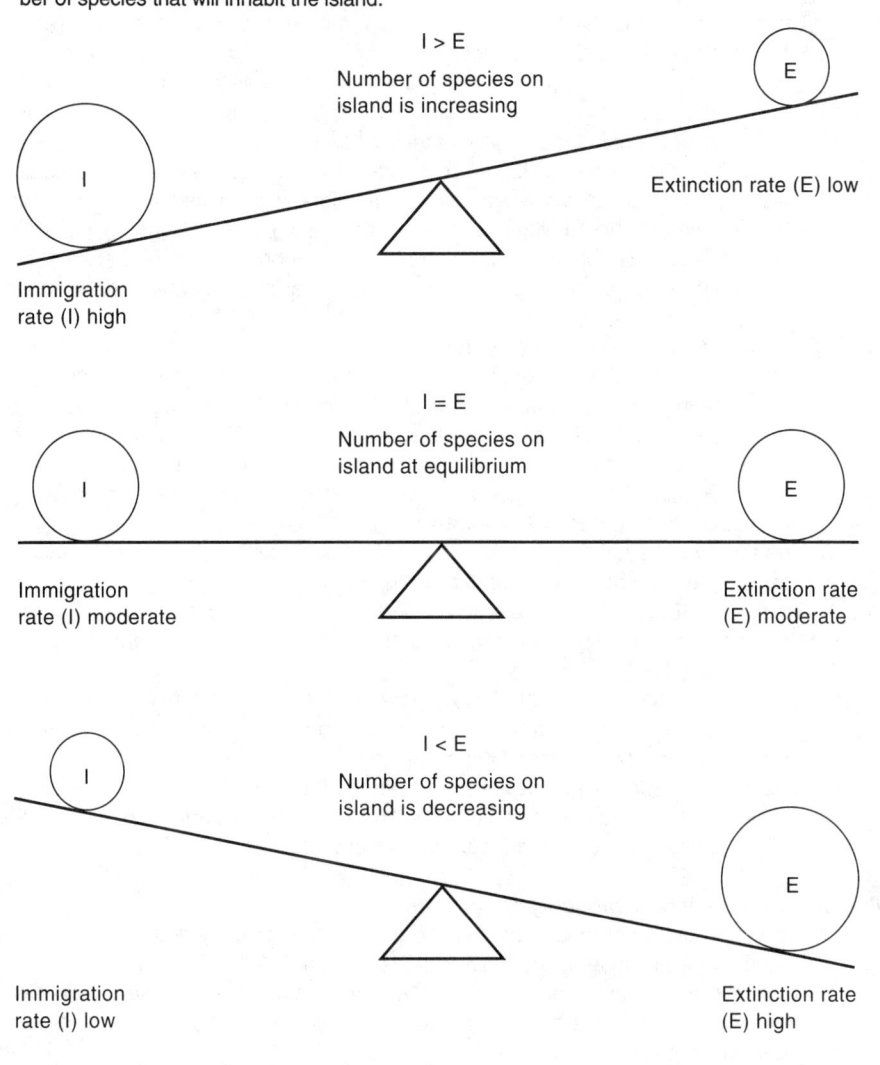

I > E
Number of species on island is increasing

Immigration rate (I) high

Extinction rate (E) low

I = E
Number of species on island at equilibrium

Immigration rate (I) moderate

Extinction rate (E) moderate

I < E
Number of species on island is decreasing

Immigration rate (I) low

Extinction rate (E) high

the populations are on isolated habitats and declines in their numbers cannot be supplemented by individuals from adjacent populations
7. The point at which the immigration rate and extinction rate balance is the equilibrium point for the developing island ecosystem and corresponds to the number of species likely to be found on the island
8. The number of species at equilibrium is influenced by the island's size and its distance from the pool of potential colonizers

a. Because smaller populations have a higher probability of extinction than larger ones and smaller islands support smaller absolute populations sizes than larger islands, the rates of extinction on small islands will be higher than those on larger islands
b. Smaller islands also present smaller "targets" for colonizers (seeds and animals) to locate
c. The farther the distance to be traveled, the fewer the species that will successfully travel that distance
d. As a result of the decline in immigration rates with distance from the source, islands close to mainland sources will tend to have higher equilibrium numbers of species than more distant islands
9. An equilibrium view of species diversity can be applied to mainland habitats as well as to island habitats when patches of habitat are isolated from other similar habitats by marked differences in terrain that impede immigration; examples of such isolated habitats include mountaintops, ponds, dunes, and areas highly disturbed by human activity, such as strip-mine tailing or clear-cut forest areas

## C. Criticisms of island biogeography theory
1. Although the island biogeography theory is valuable in understanding and investigating the role of immigration and extinction in the development of an ecosystem, empirically testing the theory has proven difficult; studies attempting to investigate the theory's efficacy have produced mixed results
2. The major criticisms of the island biogeography theory center on its simplistic approach to the processes involved
3. The theory does not adequately address the impact and importance of chance or randomness in the arrival or extinction of species
4. The theory treats all colonizers as equal, with an equal probability of extinction and immigration; this is unlikely, as some species are better adapted than others to colonize new habitats
5. The assumption that extinctions are related to island size ignores the fact that immigrations and extinctions can be related and that extinctions are influenced by the life history patterns of the organisms involved
6. Despite its shortcomings, however, the island biogeography theory has made important contributions to ecology by stimulating insights and research into the distribution, diversity, and conservation of species

## D. Application of island biogeography theory
1. One important application of the island biogeography theory relates to the management and conservation of wild plants and animals
2. With increasing human population and the attendant pressures exerted on the earth's ecosystems, a major concern is the destruction and fragmentation of natural habitats
3. Island biogeography has been applied to conservation with respect to the problems of size, shape, number, and distribution of biological reserves and preserves and the degree to which fragmentation of habitat affects species extinction
4. In general, large areas support a greater diversity of species than do small areas, and habitats closer to pools of potential colonizers contain more immigrants than do more distant habitats

5. Given these observations with regard to habitat size and location, along with the fact that large areas of habitat are being carved into smaller fragments by logging, suburban development, road construction, and other human activities, a vital question arises: Does a critical size exist for a particular habitat, below which the area will not provide the requirements of many of the original species and thus some will become extinct?
6. Based on the assumptions of island biogeography theory, certain decisions regarding the size, spacing, and arrangement of wildlife preserves can be made
   a. According to this theory, a single large preserve will protect more species than will several small preserves with the same total area, and sets of preserves should be more effective if they are located near each other and spaced more or less equidistantly rather than widely or linearly spaced
   b. Close, equidistant spacing of preserve areas should facilitate immigration among the preserve areas
   c. Smaller preserves located between larger preserves serve as "stepping stones" to facilitate movement of species between preserves
7. The suggestion that large preserves are the most desirable for conservation purposes has sparked a major controversy in conservation biology as to whether a single large preserve or several small preserves would better guarantee species survival
   a. The answer to this question depends on the areas and species under consideration
   b. In an area where the biota is more or less uniform, a single large preserve does contain more species than several small preserves
   c. In an area that contains major habitat gradients or several centers of biotic diversity resulting from historical or geological influences, several small preserves located in different parts of the area will contain more species
   d. To reduce the risk of complete extinction of an endangered species as a result of a natural event, such as flood or fire, spreading the individuals among several smaller preserves may be preferable to concentrating them in one larger preserve; however, if extinction rates are high, a single large preserve may offer better protection against extinction because population sizes are likely to be larger in larger preserves
   e. The kinds of species involved also will influence the size requirements of a biological preserve; for instance, large mammals (such as mountain lions) require larger preserves than do smaller mammals (such as rodents)

## Study Activities

1. Define and give examples of the following terms: primary succession, secondary succession, allogenic succession, autogenic succession, seral stage, pioneer stage, and climax stage.
2. Construct time lines showing the major steps of primary and secondary terrestrial succession.
3. Outline the major steps of aquatic succession from pond to terrestrial habitat.
4. Give at least one example illustrating each successional model of Connell and Slatyer.
5. List the major trends commonly observed in ecosystem attributes during succession.

# 13

# Disturbance in Ecology

## Objectives
After studying this chapter, the reader should be able to:
- Explain the role disturbance plays in the development and persistence of an ecosystem.
- Describe disturbance in terms of cause and effect on various ecosystems.
- Explain the importance of a disturbance's scale and frequency in determining its impact on an ecosystem.
- Relate the role fire plays in forest and grassland ecosystems.
- Explain the concept of ecosystem stability and its components of resistance and resilience.

## I. Disturbance in Ecosystems

### A. General information
1. Although changes in ecosystems may be undetectably slow, they do occur over time
2. Commonly, humans attempt to preserve an area by protecting it from fire, insect attack, and other perturbations considered harmful or detrimental
3. Disturbance is an important factor in maintaining diversity within ecosystems
4. In an ecological sense, a disturbance is any relatively discrete event that comes from the outside and causes changes in ecosystems, communities, populations, substrates, or resources and thereby creates opportunities for individuals or groups of individuals to become established
5. Disturbances have both spatial and temporal characteristics
   a. Spatial characteristics include the size of the area disturbed, the location of the disturbance, and its severity
   b. Temporal characteristics include the frequency of disturbances, the mean number of disturbance events per unit of time, and the mean amount of time between disturbances

### B. Agents of disturbance
1. Natural disturbances include wind, fire, moving water (storm runoff and wave action), drought, and actions of animals
   a. Strong winds associated with storm events can disturb ecosystems by causing large openings in the forest canopy ("blow-outs"), where large trees may be uprooted or otherwise heavily damaged

    b. Fires cause major changes in species composition and diversity; some ecosystems (for example, the southwestern U.S. chaparral, the African savannas, and the southern U.S. pinelands) have evolved under various fire regimes in which fire plays a major role in maintaining the community
    c. Storm runoff and flooding are major agents of disturbance that can alter the course of rivers and drastically alter the terrestrial ecosystems of the floodplain
    d. Storm waves and tides alter beach and dune structures, disturbing shore and island communities
    e. Severe or prolonged drought can dramatically alter vegetation, which, in turn, will have a major influence on animal populations in an area
    f. Overgrazing of vegetation by animals can have great impact on an ecosystem
       (1) When African elephants exceed the capacity of a particular area, their feeding habits devastate the area's flora, fauna, and soils
       (2) Insect infestations (for example, such species as gypsy moth, balsam aphid, pine bark beetle, and spruce bud worm) can defoliate large areas of forest, resulting in tree death or reduced growth
    g. Periodic flooding of forested areas in North America and Europe caused by beaver activity result in major disturbances of streams and adjacent terrestrial habitats
2. Human activities commonly result in substantial disturbance to natural ecosystems — in many cases with a more profound impact than natural disturbances
    a. Because of the tillage methods used and the monocultural aspect of cultivated plant communities, modern agricultural practices may promote erosion and the outbreak of pest species that are detrimental to adjacent natural ecosystems
    b. Surface or strip mining causes major disruptions of ecosystems as the topsoil is removed and deep unweathered rocks are brought to the surface, where rapid weathering releases toxic materials into soils and waterways
    c. The disturbance created by timber harvesting depends on the lumbering methods used; for example, clear-cutting, a lumbering practice in which all trees larger than 2.5 cm in diameter in a 20- to 40-acre forest area are removed, is the most disruptive lumbering method because it results in total destruction of the forest ecosystem
       (1) Foresters often modify the regeneration of a forest to meet their requirements by removing undesirable species, thinning out existing trees, and planting new varieties
       (2) Wholesale harvesting of a forest for the purpose of lumber production and subsequent replanting does not constitute regeneration of a forest — a forest is an ecosystem and consists of more than merely trees
       (3) The destruction of old-growth forest by the timber industry may result in permanent loss of this habitat and the species it contains
    d. Human activities have released large amounts of pollutants into the air, water, and soil that have entered natural ecosystems
       (1) Agriculture as well as road and building construction accelerate erosion and produce silt, which enters waterways and clogs rivers, streams, lakes, and estuaries

(2) The release of nutrients (such as nitrogen, phosphorus, and organic matter) from sewage and industrial effluents can accelerate eutrophication and alter the species abundance and diversity of waterways
(3) Chemicals released by agricultural, industrial, and manufacturing processes — notably pesticides, heavy metals, and toxins — have accumulated within ecosystems (bioaccumulation), causing major changes to these ecosystems

e. Reclamation of damaged or derelict land is increasingly important because of the large quantities of land that have been affected and the perceived danger they pose
  (1) Reclamation projects focus on restoring land to productivity, removing toxic substances, and restoring visual or aesthetic values
  (2) General reclamation usually involves revegetation of the site to prevent further erosion and degradation
  (3) Less emphasis is placed on reestablishing animal species, because of the belief that once vegetation is established, more mobile animals will migrate to the restored site
  (4) In urban areas, the acquisition of derelict lands for parks and open space areas is a growing trend in city planning

## C. Scale of disturbance
1. The scale of a disturbance is determined by the size of the area affected, the intensity of the disturbance, and its severity
2. The size of a disturbance must be considered in relation to the size and scale of the surrounding area; for example, the loss of a small group of trees in a small woodland parcel will have a greater impact on the ecosystem than would the loss of a similar number of trees within a large forest
3. Small-scale disturbances result when individuals or groups die and open the canopy or substrate for colonization
   a. In grasslands, the burrowing activity of badgers, groundhogs, and prairie dogs exposes patches of mineral soil for colonization by herbaceous species
   b. In forests, uprooting of a tree by wind creates a gap in the forest canopy that allows sunlight to penetrate to the forest floor and often exposes mineral soil; the result is reorganization of the vegetation
      (1) If the opening is relatively small, trees around the gap expand their crowns to fill in the opening
      (2) In larger gaps, the altered environmental conditions (that is, more light, higher temperatures, exposed mineral soil) provide opportunities for colonization by plant species different from dominant species, often altering the course of succession within the community
4. Larger-scale disturbances are induced by fire, logging, land-clearing, and other activities that involve wholesale removal of species over a large area
   a. Large-scale disturbances often result in major changes to the community and a "restarting" of succession within the disturbance site
   b. Which species colonize a disturbed area is influenced by seed availability, presence of seedlings and saplings, soil conditions, degree of competition, and other ecological factors
   c. The response to defoliation often is rapid, as seeds and seedlings of woody plants take advantage of the changed environmental conditions

## D. Frequency of disturbance
1. In temperate and tropical forest ecosystems, natural disturbances that create gaps in the canopy occur frequently but generally affect only a relatively small area of the forest
2. The slow rate of disturbance, recovery, and replacement in forests is responsible for maintaining species diversity within these ecosystems and allows many different species to coexist
3. Large-scale natural disturbances, often covering vast areas of an ecosystem, occur relatively infrequently but are of high magnitude
4. Suppression of disturbance can lead a disturbance-influenced ecosystem into a more fragile, less resilient stage in which it is more susceptible to destruction
   a. This phenomenon is seen in fire-adapted ecosystems that have evolved under a regime of periodic fires, which remove litter and combustible debris from the soil surface
   b. A large-scale fire occurring in these ecosystems often will be more damaging and destructive to the ecosystem than the smaller, more frequent fires that once occurred naturally
   c. Some plant communities depend on periodic fires to remove intruder species and allow regeneration of established species within the community

## E. Fire
1. Fires are classified as surface (litter), soil, or crown (wildfire), based on severity and the materials consumed
   a. *Surface fire,* the least severe of all fires, is the most common type in natural habitats where fires occur periodically
      (1) In a surface fire, the litter and vegetation of the forest floor are consumed
      (2) Although herbs, grasses, and small shrubs may be burned, many regrow rapidly
      (3) Many trees tend to be unaffected by fast-moving and relatively cool surface fires
      (4) Light surface fires tend to supplement the bacterial action in decomposition
   b. A *soil fire* tends to be hot and smoldering and usually results in considerable destruction of the soil's humus layer
      (1) Because of its intensity and duration, a soil fire is likely to damage plant roots and destroy seeds located within the soil and leaf litter
      (2) Soil fires occur less frequently than do surface or crown fires, because soils tend to have sufficient water to prevent burning
   c. A *crown fire* is the most dramatic and devastating fire type
      (1) Crown fires usually destroy the forest canopy, because the leaves and trunks of major forest trees become engulfed in flame
      (2) When a crown fire is coupled with heavy winds that flare it and keep it burning, the result is known as a *fire storm,* a rapidly moving fire that can reach speeds of 40 mph
      (3) Crown fires usually are damaging to most organisms because of the severe ecosystem destruction and disruption
2. Approximately 10% of forest fires are started by lightning and 90% are started by people; about 30% of all forest fires are deliberately set
3. The frequency or periodicity of fire is considered semirandom and generally depends on litter accumulation, fuel characteristics (moisture content and vol-

ume of accumulated leaves and twigs), moisture or drought conditions of soil and vegetation, and human interference
4. The impact or influence of fire on forest ecosystems can be complex and difficult to predict
   a. Temperate forests evolved in conjunction with fire, and fire historically was a recurring attribute of the ecosystem
   b. In a climax forest community, dead wood and the annual leaf or needle litter accumulate, providing fuel
   c. Periodic surface-type fires generally cause little disruption of the forest ecosystem
   d. Many tree species are long-lived; forest development (succession) brings an increase in biomass and the accumulation of stored nutrients within this biomass
   e. A fire releases this storehouse of nutrients from the biomass (both living and accumulated detritus or litter), provides nutrients to surviving plants, and often results in a period of rapid growth of these survivors
   f. The black ash residue from a fire can increase the absorption of solar radiation (causing a 3° to 16° C rise in surface temperature), heating upper soil layers and promoting bacterial action and seed germination
   g. A white ash residue suggests that considerable volatilization of nutrients (particularly nitrogen) occurred and that fewer nutrients are available for surviving species
5. A fire's effect on soil depends on the temperatures to which the soil is exposed during the fire
   a. Even with the hottest forest fires, soil temperatures rarely exceed 200° C at a 2.5-cm depth
   b. Little volatilization of organic matter from soil occurs at temperatures below 200° C, but temperatures above 200° C will often destroy 85% of the organic matter, converting it to carbon dioxide and nitrogen gas
   c. Intense heat breaks down soil aggregates and thus decreases water infiltration and soil aeration, resulting in increased runoff and erosion
   d. Fires tend to reduce fungal populations (particularly the nitrifying bacteria), lower the number of soil invertebrates, increase herbaceous legume species, and stimulate most soil bacteria populations
   e. Soil erosion and leaching of nutrients are likely to increase once the soil-litter matrix is disturbed
6. The overall effects of fire on a forest ecosystem depend on the type of forest vegetation
   a. The savanna ecosystem of the southeast United States, also known as the long-leaf pine-wire grass ecosystem, is profoundly affected by cyclic fires
      (1) Without periodic fires, fuel (dead leaves, branches, and twigs) accumulates and shrubs and scrub hardwoods invade; this inhibits the growth and survival of long-leaf pine
      (2) Suppression of fire for 4 to 5 years chokes out the long-leaf pine and leads to an overall decline in species diversity
      (3) Long-leaf pine is more resistant to fire than other tree species, because the seedlings' terminal buds are protected by a thick bunch of fire-resistant needles
      (4) In areas periodically visited by fire, species richness, reproduction (flowering), and the abundance of wind-dispersed plants all increase

(5) If the long-leaf pine-wire grass ecosystem does not burn periodically, then the habitat changes dramatically and the ecosystem is altered
   b. The lodgepole pine ecosystem extending from Colorado north to Alberta, Canada, is very well adapted to periodic fires
      (1) Lodgepole pine forms monospecific stands and its seedlings are intolerant of the conditions beneath the canopy; thus this species relies on recurring disturbances such as fire for successful reproduction and replacement
      (2) Fires appear to occur naturally at approximately 100- to 400-year cycles
      (3) Biomass accumulates for 60 to 80 years after a fire and then levels off, reaching a more or less constant level after 100 years
      (4) If fire is suppressed, the lodgepole pine will be replaced by other species capable of germinating and establishing beneath the canopy
   c. Depending upon the species of trees present and the successional age of the ecosystem, a fire often can lead to dramatic changes in a forest
      (1) Some trees (for example, aspen, paper birch, and pine) are reseeded after a fire by wind deposition, while other plants (for example, oak and perennial grasses) sprout from fire-resistant roots; the result is a potentially dramatic shift in species composition
      (2) Some tree species, such as jack pine and lodgepole pine, produce seeds that are held in the cones until the cones are heated sufficiently (from 60° C to 150° C) to release the seeds; these species therefore depend upon fire to release seeds and permit germination
   d. The ponderosa pine forests of western North America appear to burn and develop under a fire cycle occurring every 5 to 20 years
      (1) The periodic fires are of low intensity and reduce the needle layer periodically, thus preventing fuel buildup
      (2) The fires also thin the stands of pine and eliminate shade-tolerant conifer species, thereby helping the ponderosa pine to remain dominant
   e. The red pine and white pine forests of the Great Lakes region experience low-intensity fires every 20 to 38 years
7. Fire also is important for survival and maintenance of grassland ecosystems, particularly at boundaries with forests or desert ecosystems
   a. Temperate grasslands occur in regions whose climate is similar to that of temperate deciduous forests, except that water is limited
   b. Grasses have underground rhizomes, often 2.5 cm or more below the soil surface, and thus are protected from the effects of heat; in fact, the moist, living tissue above the growing point often prevents fire from destroying the meristems (plant growth centers)
   c. On the eastern boundary of the tall grass prairies of the midwestern United States, recurrent fires destroy the seedlings of encroaching tree species that would otherwise shade the grasses and out-compete them; thus, the fire favors the grasses
   d. Although the above ground portions of grasses are consumed in a fire, intercalary meristems just below the soil surface allow the grasses to resprout quickly after the fire; tree seedlings, on the other hand, generally are killed
   e. On the western edge or boundary of the grassland range, desert shrubs are kept in check by periodic fires, which open the habitat for grasses
   f. As a result of periodic fires, the grassland community is maintained and intrusion by shrubs or trees is prevented; human intervention in the form of fire

prevention may cause a dramatic shift in species composition and alter the ecosystem
8. The effects of fire on wildlife are determined by the fire's severity, the animals' mobility, and the time it takes the vegetation to recover
   a. In most fires, wildlife deaths are rare, because most animals will either hide (in burrows, dens, or trees) or flee; deaths most commonly occur in crown fires or when animals become trapped
   b. A fire's aftermath commonly brings a dramatic loss of shelter and food availability and increased exposure to predation — particularly of small rodents
   c. Short-term effects of fire on stream fauna and flora include loss of streamside vegetation (with subsequent increased sediment load), increased stream temperature, decreased oxygen content, increased algae growth, and increased numbers of insect larvae
9. Fire management or suppression can allow the accumulation of fuel, crown closure, development of an understory that acts as a fire ladder to the crown, and the aging and death of individual trees, all of which result in the potential for a more intense and devastating fire at a later date
   a. Many years of fire suppression in the forests of Yellowstone National Park prior to 1975, coupled with an unexpected drought, resulted in the large-scale fires of 1988
   b. The long-term impact of the Yellowstone fires will be a renewed and rejuvenated ecosystem

## II. Ecosystem Stability

### A. General information
1. **Ecosystem stability** is a measure of the integrity and resilience of an ecosystem that experiences a disturbance
2. Stable communities reach and maintain an equilibrium condition of a more or less steady state in which they remain and generally do not change
3. A highly stable community resists departure from equilibrium conditions, or, if directed away from the equilibrium conditions, it returns rapidly with a minor fluctuation

### B. Characteristics of stability
1. Stability has two components: resistance and resilience
   a. *Resistance* refers to a community's ability to resist change, akin to the concept in physics of inertia
   b. *Resilience* refers to a community's ability to return to its original condition after a disturbance
2. Depending on an ecosystem's response to disturbance, stability may be local or global
   a. *Local stability* is the tendency of an ecosystem to return to its original state following a minor disturbance — for example, when similar tree species fill in a gap in the forest canopy produced by wind damage
   b. *Global stability* is the tendency of an ecosystem to return to its original state following all possible disturbances or disturbances of great magnitude — for example, when chaparral returns to its original condition after a fire

c. Globally stable ecosystems exhibit low variability and high resistance to change, whereas locally stable ecosystems can be easily upset by ecological disaster (for example, oil spills, pollutant release, flash flood)
3. Forest ecosystems are relatively resistant to disturbance and can withstand such disturbances as sharp temperature changes, drought, and insect outbreaks, because the ecosystem can draw on a reserve of nutrients and energy
4. Aquatic systems exhibit little resistance to disturbance but appear to be resilient
   a. They lack any significant long-term energy or nutrient storage in biomass
   b. An influx of nutrients, as might occur in the dumping of untreated sewage into a waterway, acts as a disturbance by overloading the ecosystem with more nutrients than it can handle; because the ecosystem has very limited ability to retain the nutrients, it returns to its original condition relatively rapidly after the disturbance is removed
5. Several lines of circumstantial evidence suggest that diversity causes stability
   a. Small islands with low species diversity are much more vulnerable to invading species than are continents with higher species diversity
   b. Outbreaks of pests occur more frequently on agricultural land planted in one-crop species than in natural areas with a variety of species
   c. Pesticide overuse has caused pest outbreaks by eliminating predators and parasites from the insect community of crop plants
   d. The complexity (that is, the number of pathways through which energy may travel within an ecosystem's trophic structure) of food webs tends to be greater in stable communities than in fluctuating environments
6. Two schools of thought attempt to explain the link between stability and diversity: the equilibrium hypothesis and the nonequilibrium hypothesis
   a. The *equilibrium hypothesis* is an older, more traditional view of ecosystem organization in which communities are thought to be determined by predation, parasitism, and competition; communities are considered stable entities, and disturbances are extinguished and species diversity is determined by the number of niches available
   b. The *nonequilibrium hypothesis* is a more contemporary view in which species composition changes continuously; species composition is never in balance, and stability is rare or nonexistent
   c. Little data support one hypothesis over the other, although the nonequilibrium hypothesis is gaining support among ecologists

## Study Activities

1. List at least three natural and three human-caused disturbances of ecosystems and comment on their possible impacts.
2. Compare and contrast the different timbering techniques used in logging and give the major ecological impacts of each.
3. Explain the ecological significance of scale and frequency as it pertains to disturbance.
4. Describe the three types of forest fires and the relative impact each might have on a forest.
5. Describe and give at least four examples of fire-adapted ecosystems in which fire is essential to the community's perpetuation.
6. List the major attributes of ecological stability.

# 14

# Terrestrial Ecosystems

## Objectives

After studying this chapter, the reader should be able to:
- Describe the concept of a biome and explain its usefulness in ecology.
- Identify the nine major biomes of the world and indicate their locations and major features.
- Identify the major climatic patterns that give rise to a particular type of biome.

## I. Classification of Terrestrial Ecosystems

### A. General information
1. **Biomes** are regional ecosystems with similar communities that extend over large geographical areas; they include all animals and plants adapted to a common climate
2. Biomes are distributed in a general east-west direction on most continents and are oriented in a series of bands encircling the globe at various longitudes moving from the equator toward the poles
3. Important exceptions to the east-west distribution pattern occur in mountainous regions, where abrupt changes in altitude can cause a series of biomes along the mountainside in response to different altitudes
4. Although the general appearance of biomes may be similar over a large area, actual species composition throughout a biome varies from one location to another; thus specific biomes cannot be strictly defined

### B. The climate hypothesis
1. In 1855, plant taxonomist Alphonse de Candolle postulated that life forms of plants occurred in collections (later to be known as biomes) that were determined by the prevailing climate of an area
2. His classification consisted of five life forms, each identifiable by temperature and moisture conditions, ranging from forms of the tropical rain forest (needing much heat and humidity) to forms of the arctic tundra (needing cold temperatures and low humidity)
3. The de Candolle classification scheme has been further modified and elaborated to become the modern concept of biomes but it was surprisingly accurate in describing, within tolerable limits, the actual climate of different parts of the world

## II. Major Biomes

### A. General information
1. Major determinants of the type of terrestrial biome in a particular area are temperature and moisture
2. Ecosystems in similar climates but different geographic areas often contain similar communities
3. Biomes are classified by the major vegetation types; they include temperate grassland, chaparral, savanna, shrubland, desert, taiga, temperate deciduous forest, tropical rain forest, and tundra
4. Biomes include all animals and plants adapted to a common climate and, therefore, are environmentally linked
   a. Temperature, water, light, and wind are major components of climate or the prevailing weather conditions of a locality
   b. Climatic conditions, particularly moisture and temperature conditions, have a great impact on the distribution of biomes

### B. Temperate grassland
1. *Temperate grasslands* are located in temperate areas of South Africa, Argentina, North America, and southeastern Europe and Asia
2. They have some characteristics of the savanna (tropical grasslands) but occur in regions with colder temperatures
3. Grasslands are perpetuated by periodic fires and droughts, which prevent colonization and permanent establishment of trees and shrubs
4. They tend to integrate with forests, woodlands, and deserts at their margins
5. Their soils tend to be deep and rich in nutrients
6. Grassland animals include numerous birds, rodents, large herbivores (wild horse, buffalo), and large predators (wolves, coyotes, leopards, cheetahs, hyenas, lions)

### C. Chaparral
1. *Chaparrals* are regions of dense, spiny shrubs with tough evergreen leaves
2. They are located in mid-latitude areas along the coasts of California, Chile, southwestern Africa, and southwestern Australia
3. Fire plays an important role in maintaining the characteristic vegetation of the chaparrals
   a. Chaparral shrubs have root systems adapted to fire or produce seeds that germinate after being subjected to the heat of a fire
   b. Other competitor plants such as grasses and trees cannot recover as readily from a fire, and so the shrubs persist
4. Animals of the chaparral include deer, fruit-eating birds, rodents, snakes, and lizards

### D. Savanna
1. *Savannas* are considered tropical grasslands with scattered, individual trees
2. They are located in central South America, central and south Africa, and parts of Australia
3. Savannas generally are tropical and subtropical regions with three seasons: cool and dry, hot and dry, and warm and wet
4. Fires frequently occur, started by such natural causes as lightning

a. Many of the grass species can regrow rapidly after a fire
b. The trees of the savanna generally are not very fire-tolerant; thus fires prevent trees from invading and establishing a forest
5. Trees of the savanna generally are deciduous and lose their leaves during the dry seasons
6. Animals of the savanna include large herbivores (giraffe, zebra, antelope, buffalo, kangaroo), snakes, rodents, and many arthropods

## E. Shrubland
1. *Shrublands* are naturally found in arid and semiarid regions; they also occur in temperate regions where human activities have affected succession and prevent development of forest vegetation
2. Shrubs are plants with woody persistent stems, no central stem or trunk, and a height of less than 6 meters
3. They survive in areas where trees cannot because they have less investment in aboveground structures and have structural modifications to dissipate heat and reduce evaporative losses
4. Ecologists recognize four types of shrublands, based primarily on location and characteristic plants: Mediterranean, northern desert, heathland, and successional
    a. Mediterranean shrublands occur in the semiarid regions of western North America, regions bordering the Mediterranean Sea, central Chile, the Cape region of South Africa, and the southwestern and southern regions of Australia
        (1) The Mediterranean climate is marked by hot, dry summers with at least one month of prolonged drought and cool, moist winters
        (2) All of the Mediterranean shrublands support xeric (low-moisture) broadleaf evergreen shrubs and dwarf trees with some minor herbaceous understory
        (3) In general, Mediterranean shrublands lack an extensive understory and ground litter, produce large numbers of seeds, and are prone to periodic fires
        (4) Some of the species require exposure to fire to germinate
    b. Northern desert shrublands are found in the cool, arid regions east of the Rocky Mountains
        (1) The climate is continental, with characteristically warm summers and prolonged cold winters
        (2) Major vegetation includes the sagebrush and shad scale
        (3) Some authorities classify the northern desert shrubland as desert, but the vegetation type and animals are different
    c. Heathlands are found in Europe, northern Eurasia, North America, India, southeastern Asia, and the Malaysian Archipelago
        (1) Heathland vegetation is a combination of evergreen shrubs and bushes that generally have thick, leathery leaves and are adapted to fire
        (2) Heathlands are characteristic of cool to cold climates
        (3) Depending on precipitation level, the heathland will be either dry heath or wet heath
            (a) Dry heathlands occur on well-drained soils and are subject to seasonal drought

(b) Wet heathlands are subject to periodic waterlogging and characteristically contain a greater percentage of grasses and sedges than do dry heaths
  d. Successional shrublands occur on drier upland areas or in wetter areas along streams and around lakes
    (1) Successional shrublands can be considered extended, prolonged successional stages that may persist for extended periods
    (2) In drier regions the shrubs are scattered or clumped in grassy fields, and the open areas contain seedlings of forest trees
    (3) In these drier areas, sumacs, chokecherry, hazelnut, and dogwoods form hickets or dense patches
    (4) In the wet areas along water bodies, the successional shrubland forms an intermediate habitat between the meadow and the forest
    (5) Typical wet thicket shrubland plants include alder, willow, and red osier dogwood
    (6) These shrubland thickets are important food and cover areas for wildlife species

**F. Desert**
  1. *Deserts* occur wherever precipitation is consistently low or the soil is too porous to retain water
  2. Most deserts receive less than 12.5 cm of precipitation per year; thus, evaporation generally exceeds rainfall
  3. Most deserts are characterized by large shifts in daily temperature: air temperatures can extend above 36° C during the day and drop below 15° C at night
  4. Deserts may be classified as either hot or cold, depending on the prevailing temperature
    a. Hot deserts are found in North America, South America, northern Africa, the Middle East, and Australia
    b. Cold deserts are located in North America, Argentina, and central Asia
  5. The life cycles of desert plants are keyed to rainfall, not temperature, as evidenced by the large floral blooms that occur immediately after infrequent rainstorms
  6. Light intensities are high in deserts; many desert plants are adapted to these higher intensities through crassulacean acid metabolism (CAM) photosynthesis
    a. In CAM plants certain organic acids accumulate at night in plant tissue
    b. These organic acids are converted to carbon dioxide during the day
    c. CAM plants open their stomata at night to allow gas exchange and close them during the day to prevent excessive water loss
    d. CAM metabolism permits more photosynthesis than would otherwise be possible
  7. Desert plants typically possess certain attributes to survive dry conditions
    a. Many desert plants are annuals and avoid drought by growing only when there is rain, spending the dry seasons as seed
    b. Other desert plants are succulents and store water in modified roots or stem structures
    c. Desert shrubs have short trunks, numerous branches, and small, thick leaves that may be shed during prolonged dry spells
    d. Most desert plants have sharp spines or aromatic odors and chemical defenses to protect them against water-seeking herbivores

e. Desert plants generally are widely spaced to accommodate large root systems that allow maximum water-gathering capacity within the thin, porous soils
8. Desert animals typically are adapted to activity at night when lower temperatures prevail; species include rodents, reptiles (snakes and lizards), numerous insects, and a variety of birds

## G. Taiga (northern coniferous or boreal forest)
1. The *taiga* extends across North America, Europe, and Asia and is found north of the broad-leaved temperate forest and south of the arctic tundra
2. Its climate is characterized by harsh winters (with temperatures dropping as low as $-50°$ C) and brief but warm summers (with highs ranging from $20°$ to $30°$ C)
3. Precipitation usually is considerable; most occurs in the summer
4. Snow covers the area during the winter
5. Taiga soils tend to be thin, acidic, and low in mineral nutrients but often have an accumulation of litter
6. Soils in the northern portions of the taiga are underlain by permafrost
7. Conifer species of the northern boreal forest typically include spruce, pine, fir, and hemlock
8. Animals of the taiga include rodents, large herbivores (caribou), and various birds

## H. Temperate deciduous forest
1. *Temperate deciduous forests* are located in mid-latitude regions with sufficient moisture to support large trees — for example, the eastern United States, most of western Europe, and parts of eastern Asia
2. Temperate forests are characterized by broad-leaved deciduous trees
3. Temperatures range from very cold (below $-30°$ C) in winter to hot (above $30°$ C) in summer
4. Precipitation is relatively abundant and evenly distributed throughout the year, with slightly more in the summer
5. During the summer, the trees form a relatively closed canopy that shades the forest floor from direct sunlight
    a. This shading effect results in a progression of flowering through the growing season
    b. In the spring before the canopy is fully formed, spring flowers bloom on the forest floor
    c. Other, more shade-tolerant plants bloom later in the season, typically in mid-summer through fall
6. Animals of the temperate deciduous forest include rodents, snakes, amphibians, and a variety of birds

## I. Tropical rain forest
1. *Tropical rain forests* are located near the equator where day length remains steady, providing approximately 12 hours of light per day throughout the year
2. Temperature in the rain forest also varies little year round, generally ranging between $25°$ and $35°$ C
3. Rainfall is the major determinant of vegetation and normally ranges between 200 and 400 cm per year
4. Although rainfall can differ from month to month, most tropical rain forests have no dry season — a condition that may be termed nonseasonal

5. High rainfall and constant hot temperatures support a great diversity of flora and fauna; as a result, rain forests contain more species than all the other biomes combined
6. Competition for light is a strong selective force in many rain forests, because the canopy becomes very dense and blocks most light from reaching the forest floor
7. Many of the trees are covered with plants, such as orchids and bromeliads
8. Despite the abundant vegetation, little accumulation of litter or humus occurs, and the soils generally are thin, low in nutrients, and acidic
9. Decomposition occurs rapidly, and the nutrients from any decomposing materials are quickly recycled into the plant community or leached away by the heavy rains
10. Animals of the tropical rain forest include primates, large predators (tigers, jaguars), numerous insects, and a large variety of birds

## J. Tundra
1. The *tundra* marks the northernmost limit to plant growth; it comprises approximately 20% of the earth's land surface
2. Plant forms are limited to low shrubby or matlike vegetation because of cold temperatures and scouring action of persistent winds
3. The Arctic tundra represents the northernmost limit of plant growth and extends from the northern limits of land to the northern edge of the taiga; the Alpine tundra is located above the tree line on high mountains
4. Arctic tundra soils are thin and underlain by *permafrost,* a constantly frozen ground layer that restricts root growth and water movement within the soil
5. Vegetation of the tundra exhibits slow rates of photosynthesis all year round
6. Dominant forms of tundra vegetation are shrubs, sedges, grasses, mosses, and lichens
7. Miniature willow and birch trees grow, although they are very prostrate, reaching heights of only 2.5 to 50 cm
8. Tundras have a brief growing season of 2 to 3 months (although frost can occur on any day of the year) marked by nearly 24 hours of daylight, during which perennial plants often form brilliant flowering mats in meadow areas
9. Because of the harsh environment and thin topsoil layer (only 5.0 to 7.5 cm thick), the tundra tends to be very fragile, and any disturbance is likely to remain for long periods
10. Animals of the tundra include rodents, predators (arctic fox, snowy owl, polar bear), caribou, and birds

# Study Activities
1. Construct a table listing the major biomes of the world along with their major characteristics.
2. Describe the biome or biomes located within your state or locality.
3. List the biomes in which fire plays a major role.
4. Explain the climate hypothesis of plant life forms as first devised by Alphonse de Candolle and how it relates to the modern concept of biomes.
5. Choose two biome types and describe their similarities and differences.

# 15

# Freshwater Ecosystems

## Objectives

After studying this chapter, the reader should be able to:
- Explain water's unique chemical and physical properties that influence aquatic ecosystems.
- Describe the basis of water's biologically important properties.
- Distinguish the major features of lotic and lentic ecosystems.
- Identify and describe the major biological regions of a lake or pond.
- Describe the major organisms found in lotic and lentic ecosystems.
- Explain the important features of wetlands and the different types of wetland habitats.
- Describe the ecological role and importance of wetlands.

## I. Physical and Chemical Characteristics

### A. General information
1. Many environmental parameters important to aquatic ecosystems arise from water's molecular structure and its resultant physical and chemical properties
2. Chemically, water is a polar molecule that readily forms hydrogen bonds with adjacent molecules
3. Hydrogen bonding between water molecules is the basis for many of water's unique properties
4. Water's biologically significant properties include adhesion, cohesion, surface tension, high specific heat, high heat of vaporization, versatility as a solvent, and the lower density of ice compared to liquid water

### B. Interaction of temperature, density, and salinity
1. Water's *density* (mass per unit volume) changes with temperature
   a. Density increases as temperature declines and is greatest at 4° C; when water freezes, however, density decreases
   b. As density increases, organisms must expend greater amounts of energy to move through water
2. *Salinity* is a measure of the quantity of dissolved salts in water
   a. Seawater contains 35 or more parts per thousand of dissolved salts
   b. Freshwater contains approximately 0.5 parts per thousand of dissolved salts
   c. *Conductivity,* the total quantity of dissolved ions in water, measures water's ability to carry an electric current

d. Deionized water has a conductivity of zero; higher conductivities suggest greater numbers of dissolved ions
   e. In addition to sodium and chloride ions in the form of salt (sodium chloride), other ions typically found in natural waters include calcium, magnesium, and potassium
   f. Dissolved ions are important to aquatic organisms, because the presence and concentration of ions determine osmotic relationships — whether the organisms will gain or lose water to their surroundings by diffusion
 3. Waters of different temperature and salinity will have different densities
   a. Waters of different densities do not readily mix
   b. The result is *stratification,* the formation of different layers of water of varying densities

## C. Light penetration
 1. Sunlight penetrating a water body attenuates with depth; it is quickly absorbed by water molecules and by materials dissolved or suspended in the water column and converted to heat
 2. Different wavelengths (or colors) of light are absorbed differently by water
   a. Red light is most strongly absorbed and penetrates the water column to only a shallow depth
   b. Blue light and green light are less strongly absorbed and penetrate to a greater depth
 3. The penetration of various wavelengths is an important determinant of photosynthesis; red and blue wavelengths are most effective for photosynthesis and green light, although it penetrates to the greatest depth, is least effective
 4. Very turbid or cloudy waters will attenuate light quickly and severely limit photosynthesis
 5. The depth at which the energy harnessed by photosynthesis equals the energy expended by respiration is known as the **compensation point**
   a. The compensation point occurs at the depth where 1% of the original incident light penetrates
   b. The region above the compensation point, where the net energy gain from photosynthesis exceeds respiration, is called the euphotic zone
   c. The region below the compensation point, where respiration exceeds photosynthesis, is termed the profundal zone

## D. Oxygen concentration
 1. Oxygen enters a water body by diffusion from the atmosphere at the water surface or as the result of photosynthesis in aquatic plants and algae
 2. Oxygen concentration in a water body is affected by temperature, salinity, and altitude
   a. Cooler water can hold more oxygen than warmer water
   b. Water's oxygen-holding ability decreases with increasing salinity
   c. Generally, higher altitude results in lower atmospheric oxygen concentrations and, consequently, lower oxygen concentrations in water bodies
 3. Oxygen concentration can be measured directly in terms of parts per million and also expressed as the percent saturation, or the amount actually present as compared to the theoretical level

## 142  Freshwater Ecosystems

4. Another related and important measure of oxygen, **biochemical** or **biological oxygen demand (BOD),** is a measure of the amount of oxygen consumed by organisms in a water sample
   a. BOD is determined by measuring oxygen concentration in a water sample kept in a dark room at 20° C before and after a 5-day period; the difference in oxygen concentration at the end of the test period is the BOD of the sample
   b. High BOD generally indicates high microbial activity and often suggests water pollution

### E. Carbon dioxide and pH
1. The carbon dioxide concentration in a water body is related to the balance of photosynthesis and respiration within the ecosystem
2. Carbon dioxide is approximately 200 times more soluble in water than oxygen; thus, carbon dioxide concentrations may reach much higher levels than oxygen concentrations
3. In addition to simply dissolving in water as a gas, carbon dioxide also will chemically react with water to form bicarbonate ($HCO_3^-$), carbonate ($CO_3^{-2}$), hydrogen ($H^+$), and hydroxide ions ($OH^-$)
4. The abundance of hydrogen and hydroxide ions determines the water's pH
   a. The *pH* of a solution is defined as the negative logarithm of the hydrogen ion concentration, expressed in moles per liter
   b. The pH scale ranges from 0 to 14; solutions with pH below 7 are considered acidic, those with a pH of 7 are neutral, and those with a pH above 7 are alkaline, or basic
5. Natural waters contain various materials that interact with hydrogen ions to alter the ions' effect on the water's overall acidity, a process known as *buffering*
   a. The concentration of calcium and magnesium ions combined with either bicarbonate or carbonate ions is termed *hardness*
   b. As water's hardness increases, so does its buffering capacity
   c. Generally, hard freshwater and seawater are highly buffered because of the high concentration of dissolved ions
   d. Soft freshwater has few dissolved ions and consequently has a lower buffering capacity
6. Aquatic plants and photosynthetic algae can consume dissolved carbon dioxide as well as bicarbonate ions (as carbon supplies for photosynthesis) and thereby often affect water pH
   a. During daylight hours, the carbon dioxide and bicarbonate ion supply may be exhausted in some waters; the result may be increased pH during the daytime
   b. At dark, when photosynthesis ceases and respiration predominates, pH will drop

### F. Water movement
1. Water movement within a water body has a great impact on the type of ecosystem that develops
2. Currents may be caused by gravity, wind, density differences, and temperature gradients
   a. Currents are important in mixing the water column, in distributing nutrients and other materials, and in distributing organisms

b. The velocity of water flow also influences the degree of siltation and patterns of silt deposition
3. Surface waves on a water body are caused by the action of the wind on the water surface
   a. Except for their effects on the shoreline and shore organisms, surface waves generally are not biologically important
   b. In a surface wave, each water molecule remains largely in the same place with respect to lateral movement and generally follows an elliptical orbit vertically in the water column as the wave passes
   c. As waves approach land, they encounter shallower waters, and their height increases above the water surface; they eventually grow too tall and topple over, thus resulting in the shifting wave action at the shoreline
4. As the wind blows across a large, open water body (lake or pond), it drives water to the leeward end (in effect, piling the water up on the leeward side) and creates a depression on the windward edge
   a. The overall effect of the wind distributing the water across the water body is rocking or "sloshing" within the lake basin, termed a *seiche*
   b. A seiche can cause mixing of water within the water column and thus have an impact on the ecosystem

## II. Overview of Lotic and Lentic Systems

### A. General information
1. Aquatic ecosystems can be classified based on the type of water current
   a. *Lentic* systems have slow water movement, and the water tends to be deep and standing; these systems include ponds and lakes
   b. *Lotic* systems have strong water currents or movements (flowing water); they include rivers and streams
2. Lentic and lotic systems are connected to one another and to the oceans, atmosphere, and surrounding landscape through the hydrologic cycle (see Chapter 5, Movement of Materials Through Ecosystems)

### B. Comparison of lotic and lentic systems
1. The velocity and strength of the water current is a controlling and limiting factor in lotic systems but is far less important in lentic systems
2. Land and water exchange is very important in the lotic ecosystems and in the littoral zone of lentic ecosystems but is less important in the limnetic zone
3. Oxygen usually is abundant or near saturation in lotic ecosystems because of riffles and the churning of flowing water

## III. Lentic Systems

### A. General information
1. The lakes and ponds that characterize lentic systems are inland depressions containing standing water and can range in size from a few acres to thousands of square miles; the distinction between a lake and a pond generally is one of size and bottom substrate

## Lake Structure

Lakes and ponds can be divided into different zones based on differences in light penetration, kinds of organisms, and features of the habitat.

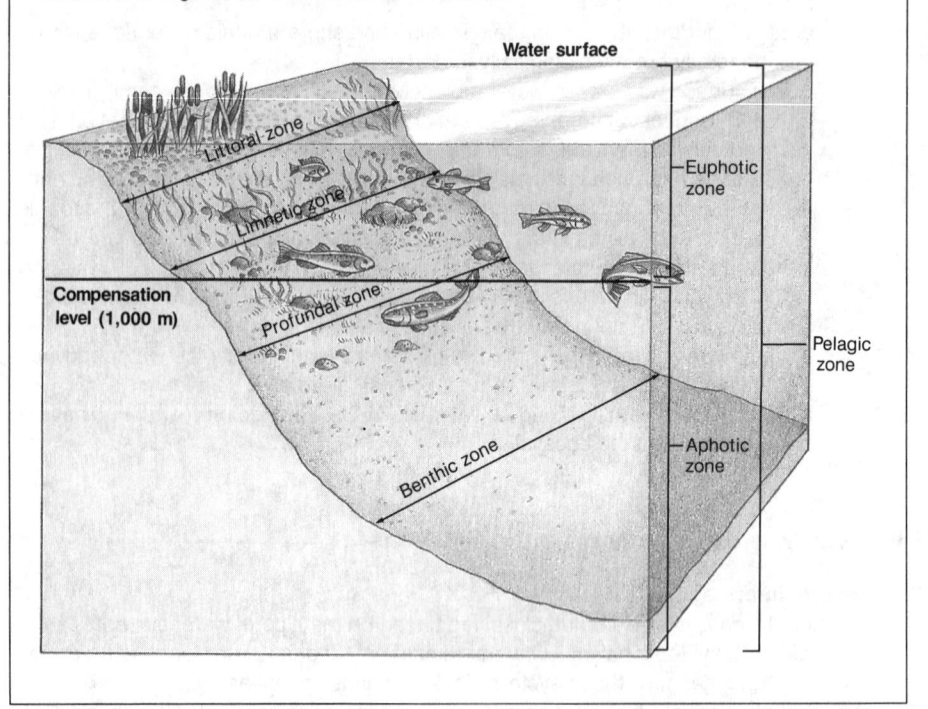

*[Diagram showing a cross-section of a lake with the following labeled zones: Water surface, Littoral zone, Limnetic zone, Profundal zone, Benthic zone, Euphotic zone, Pelagic zone, Aphotic zone, and Compensation level (1,000 m)]*

    a. Lakes tend to be larger and deeper than ponds; they may have depths to which light cannot penetrate

    b. Ponds tend to be small and shallow, with muddy or silty bottoms; generally, light penetrates to the bottom at all depths

  2. Although largely self-contained, lentic ecosystems are strongly influenced by input and outflow of nutrients and materials from sources outside the basin; such input and outflow may come from precipitation, runoff, stream flow, evaporation, and migration of animals

**B. Structure**

  1. Lakes and ponds can be divided into different zones based on differences in light penetration, kinds of organisms, and certain habitat features (see *Lake Structure*)

  2. The ***limnetic zone,*** the upper region of open water area, is dominated by the plankton community, with phytoplankton serving as the producers in the food chain

  3. The ***littoral zone,*** the region nearest shore where sunlight penetrates to the bottom, generally contains rooted plants and has the greatest concentration of organisms; the emergent vegetation in the littoral zone is an important link between land and water

  4. The ***profundal*** or ***sublittoral zone*** lies below the depth of light penetration and lacks producers and thus must rely on the importation of energy and materials

from the littoral and limnetic zones; the diversity and abundance of life in this zone are influenced by the availability of oxygen and organic material and temperature
   5. The **benthic zone** is the lake bottom; dominant organisms include anaerobic bacteria and periphyton
   6. The **euphotic zone** is a general term for all upper regions of a water body from the surface to the compensation point, including both the littoral and limnetic zones, that receive light
   7. The **aphotic zone** (meaning "zone without light") is the general term for all of the lower regions of a water body below the compensation point to the bottom
   8. The **pelagic zone** comprises both the euphotic and profundal zones and includes both the epilimnion and hypolimnion (described below)

**C. Temperature zones**
   1. As light penetrates the water column, it is absorbed and either used in photosynthesis or converted to heat; the resultant uneven heating of the water column leads to its layering or stratification
   2. On the basis of relative temperature, the water column may be divided into three main parts: epilimnion, metalimnion, and hypolimnion
   3. The **epilimnion,** the upper region of the water column, is heated by solar radiation; it tends to be high in oxygen as a result of the abundant photosynthesis of phytoplankton
   4. The **metalimnion,** the region between the upper epilimnion and the lower hypolimnion, is characterized by an abrupt drop in temperature with increasing depth, known as the **thermocline**
   5. The **hypolimnion,** the lower region of the water column, contains colder water that typically is low in oxygen but relatively high in dissolved and suspended organic matter
   6. If the thermocline lies below the compensation point, which is common, then the hypolimnion becomes oxygen-depleted

**D. Seasonal changes in the water column**
   1. Lakes and some ponds in temperate climates often exhibit a seasonal pattern of stratification caused by differences in temperature and photosynthesis; this, in turn, may influence oxygen and nutrient distribution
   2. As the summer progresses, the epilimnion becomes nutrient-poor as a result of algae growth, and the hypolimnion is oxygen-depleted because of the accumulation of decaying organic debris filtering down through the water column from the epilimnion
   3. In autumn, the epilimnion cools and mixes with the hypolimnion, causing the *fall turnover*
   4. During the winter, ice may form at the surface and become covered with snow; this cover decreases light penetration into the water column and decreases productivity
   5. In the spring, the epilimnion is warmed and again mixes with the hypolimnion in a process known as *spring turnover*
   6. The hypolimnion's degree of oxygen-depletion depends on the amount of decaying material in the water and the depth of the thermocline; productive lakes are subject to greater oxygen depletion than oligotrophic lakes as a result of in-

creased settling of detritus materials, which filter into the hypolimnion's profundal zone
  7. During the spring and fall turnover periods, the lake's upper layers receive nutrients from the deeper waters of the hypolimnion; this influx of nutrients may result in rapid growth of algae, termed an *algal bloom* (a population explosion of algae)

## E. Types of organisms
  1. Aquatic organisms are categorized based on their life forms and location within the water body
  2. Algae and protozoans (single-celled microscopic or near-microscopic organisms) that cling to plants, wood, or rocks in the littoral zone are called **periphyton**
  3. The small plants (phytoplankton), protozoans, and animals (zooplankton, such as rotifers, copepods, and small crustaceans) that float or weakly swim with the currents in the limnetic and littoral zones are called **plankton**
  4. Organisms found at or on the water surface (for example, the water strider, whirligig beetle, and mosquito larvae) are known as **neuston**
  5. Free-swimming and diving organisms found in the littoral and limnetic zones are called **nekton;** some species, such as frogs and salamanders, are confined to the littoral zone
  6. Animals that dwell on the lake bottom in the benthic zone (such as crayfish, worms, insect larvae, mollusks, and bacteria) are called **benthos**
  7. Organisms living in the sediment at lake bottom are known as **psammon**

## F. Nutrient status
  1. The nutrient status of lentic systems is tied to the relationship between the land and the water; most nutrients in lakes come from precipitation, groundwater input, and biological sources
      a. *Limiting nutrients* restrict or limit algal and plant growth when they are absent or are present in insufficient quantity
      b. In most lakes, nitrogen and phosphorus are the limiting nutrients
      c. When nitrogen or phosphorus levels increase through activities such as farming, shoreline erosion, septic system failure, or influx of sewage effluent, excessive algae growth can result
      d. Total phosphorus concentrations exceeding 30 micrograms per liter (also called parts per billion) indicate enriched conditions and likely will result in substantial algal growth
      e. Nitrogen is less likely to be a limiting factor, because it can be supplied through atmospheric contact; concentrations generally are less than 1 milligram per liter (part per million) in most lakes not experiencing explosive algal growth
  2. On the basis of successional age and nutrient content, lentic systems can be considered oligotrophic, mesotrophic, eutrophic, dystrophic, or marl
  3. **Oligotrophic** lakes and ponds are successionally young, have a low nutrient content (usually less than 10 parts per billion of phosphorus), low surface to volume ratio, clear water, and low productivity
  4. **Mesotrophic** lakes are in the intermediate stages of succession between oligotrophic and eutrophic; phosphorus concentrations generally range from 10 to 26 parts per billion

5. As the lake or pond basin fills in, the lake becomes *eutrophic;* this type is characterized by increased nutrient availability (usually high in nitrogen — above 1 part per million — and phosphorus — above 26 parts per billion), high levels of sediment and particulate matter in the water column, high surface to volume ratio, heavy algae and aquatic plant growth, and reduced light penetration caused by particulate matter and algal growth
6. As a lake progresses from the oligotrophic to the eutrophic stage, the species numbers decline, although the numbers and biomass of organisms may remain high
7. *Dystrophic* lakes receive large amounts of organic matter from surrounding watersheds; generally, they have brown water and low phytoplankton productivity but highly productive littoral zones, and often support bog vegetation along their edges
8. Although nearly all lakes are thought to naturally progress from oligotrophic to eutrophic stages, human activity (such as fertilizer runoff from agricultural land or sewage input) can speed the eutrophication process
9. *Marl* lakes are those containing extremely hard water caused by high calcium concentrations; generally, they are unproductive because of reduced nutrient availability
10. These lakes often are supersaturated with calcium and carbonates, which can react chemically with organic matter to form marl deposits; these deposits may accumulate on the lake bottom in sufficient quantity to fill in the lake and cause the development of a bog

## IV. Lotic Systems

### A. General information
1. Lotic systems include rivers, streams, and related habitats
2. One of the most important physical features of lotic systems is the strong, unidirectional current or flow downstream
3. As the water currents change speed and direction, patterns of erosion and deposition alter the habitat and may alter the stream's path
4. Because of differences in size, volume, and flow, streams and rivers have some important similarities and differences
    a. Streams generally have oxygen levels near saturation, cooler temperatures, large organic input from streamside, large fluctuation in water level, shallow water depth, a rock or gravel bottom, and a narrow channel
    b. Rivers tend to demonstrate stratification with respect to oxygen levels and have warmer temperatures; nutrient input from upstream tributaries; more stable water levels; greater water depth; a silt, sand, or clay bottom; and a broad channel

### B. Nature of lotic habitats
1. Rivers, streams, and stream banks are intimately related to the surrounding landform or landscape
2. The vegetation of the stream bank is known as *riparian*
3. Rivers may form *floodplains* — broad, flat areas adjacent to rivers that periodically flood and receive deposition of sediment from the river

4. River channels generally are winding; erosion occurs on the outer banks of bends, and deposition occurs along the inner banks of curves
5. Channels have a pattern of alternating riffles and pools
   a. *Riffles* are shallow areas with coarse substrate, high water velocity, and turbulent flow
   b. *Pools* are deeper areas with finer substrate, lower water velocity, and a smoother, more laminar flow; they may form at bends of streams or in straight runs for reasons not well understood
   c. In most streams, pools are spaced at intervals of four to seven times the channel width; pools tend to be self-sustaining
6. The amount of sediment and debris carried by a stream or river is influenced by velocity of water flow; high velocity flow carries large amounts of debris and sediment, whereas low velocity flow carries smaller particles and results in deposition
7. Water chemistry of rivers and streams tends to change more rapidly than in lentic systems because of the faster runoff from surrounding banks and the length of the banks
8. Plankton are not as common in lotic as in lentic systems, but they do occur in slower-moving areas or in pools
9. Many of the animals in streams (for example, insect larvae, snails, crayfish) are part of the benthos

## C. The river continuum
1. Because water constantly is flowing from tributaries to river mouth, a continuum or spectrum of habitats gradually intergrades from upper stream tributaries to the lower river mouth
2. The upper stream region, near the tributary origin, is swift, cold, and generally located in forested regions
   a. The riparian vegetation on either side of the forested stream reduces light and contributes more than 90% of organic input
   b. In the upper stream region, dominant organisms are shredders and collectors of vegetation within the stream, populations of grazers are minimal, and the major predators are small fish
   c. Headwater streams are accumulators, processors, and transporters of particulate matter from terrestrial ecosystems
   d. Organisms are adapted to a narrow temperature range, reduced nutrient regime, and maintenance of their position in the current
3. Further downstream, the small streams and headwaters join to form medium-sized creeks and rivers
   a. In these medium-sized streams, the importance of riparian vegetation and detrital input declines
   b. Because the streams now are wider, they lack shading in the channel center; this results in higher water temperatures
   c. The current generally slows and pools, and riffle areas develop, resulting in a greater diversity of microhabitats
   d. Dominant consumers are plant grazers and collectors, which feed on material washing down from headwaters
   e. Predators show little increase in biomass as compared to those in upstream areas, but a shift from cold water species to warm water species occurs
4. Riverine conditions develop as the channel widens and deepens

a. As the volume of water flow increases, current velocity slows and sediments accumulate on the bottom
b. The slower, deeper water and greater dissolved organic matter support phytoplankton and associated zooplankton populations

## V. Freshwater Wetlands

### A. General information
1. Wetlands are transitional lands found between terrestrial and aquatic systems where the water table usually is at or near the surface
2. Wetlands generally have one or a combination of the following characteristics: *hydric* (undrained) soils, a nonsoil substrate often saturated or covered with water at some time during the growing season, and a substrate that periodically supports predominately hydrophytes (aquatic plants)
3. Wetlands may be permanently or seasonally wet
4. Wetlands found along the edge of ponds and lakes probably represent the final stages of eutrophication in the later stages of succession in aquatic ecosystems

### B. Marshes, peatlands, and swamps
1. *Marshes* are wetlands dominated by emergent vegetation — plants with roots in soil covered by water and leaves above water
2. *Peatlands* or *bogs* are wetlands with considerable amounts of water retained by an accumulation of partially decayed organic matter
3. *Swamps* are wooded wetlands and generally are found in river valleys and along water courses
4. Freshwater marshes and swamps generally have a fluctuating water table; marshes typically have more continuously flooded conditions than do swamps
   a. The vegetation of marshes and swamps generally is tolerant of cyclic or seasonal inundation
   b. Marsh vegetation is dominated by reeds, sedges, grasses, and cattails
   c. Swamps may be classified according to water depth and location
      (1) *Deep water swamps* are found along the floodplains of larger southern rivers and on upland coastal plains and are dominated by cypress, swamp oak, and tupelo
      (2) *Upland swamps,* which may occur in various areas, are shallow water swamps dominated by willows, alders, and other shrublike vegetation
   d. Trees must cope with the problem of anchoring in the soft and unsupported soils by producing massive root systems and proplike buttressing roots
   e. The variations in environmental conditions within the marshes and swamps provide a diversity of microhabitats for such animals as turtles, snakes, alligators, and herons and even such terrestrial species as deer and bear
5. Peatlands or bogs develop in areas of blocked drainage and have cushionlike vegetation; most have some marginal semifloating mat of vegetation and usually some sphagnum moss, heaths, evergreen trees, and shrubs
   a. Organic matter production by living organisms exceeds the rate at which the compounds are respired and degraded; the net result is an accumulation of partially decayed plant and animal tissue known as *peat*
   b. Bogs most commonly are associated with northern climatic regions
   c. Lakes, ponds, and streams often are associated with bogs

d. Because drainage is inadequate, the area is saturated with water, and little organic matter is carried away or mixed with mineral substrate
e. Over time, the drainage is further congested by plant growth and the circulation of water in these areas is reduced, creating anaerobic conditions that further slow decomposition and promote the accumulation of peat
f. Bogs are stressful environments as a result of high acidity and few nutrients
g. The predominant vegetation in many bogs is sphagnum moss, which serves as a natural "sponge," holding great amounts of water and gradually releasing it during drier parts of the year; for this reason, bogs often are at the headwaters of many streams and help control stream flow
h. Animal life is restricted to a relatively few species with large populations, such as insects, rotifers, protozoa, zooplankton, birds, and rodents

## C. Importance of wetlands
1. Wetlands are valuable ecosystems because they perform a number of vital ecological functions, some of which are beneficial to humans
2. Wetlands are integral parts of watersheds because they serve as collection basins for runoff, filtering and storing the water
3. Wetlands serve in flood control by providing water storage, reducing flood peaks, slowing flood waters, and evening out the duration of flow
4. Because some wetland plants can remove nutrients and pollutants from water as it passes through, wetlands are important in maintaining water quality
5. Wetland vegetation can dampen wave action, anchor shorelines, and abate erosion through the binding action of roots
6. The primary production of fresh and tidal water wetlands ranges from 2,000 to 2,500 grams per square meter and is equal to that of many tropical ecosystems; the biomass produced in the wetlands provides a substantial base for many aquatic and terrestrial food chains
7. Because numerous denitrifying bacteria reside in wetland habitats, wetlands are important in the global nitrogen cycle
8. Wetlands serve as "living laboratories" for education and public awareness because of the diversity of plant and animal life present
9. Because of the acidic and anaerobic conditions within many wetlands, a pollen record is preserved in their sediments, which can reveal the history of regional vegetational and climatic changes in an area dating back nearly 12,000 years
10. Gases produced and released into the atmosphere by wetlands may play a role in maintenance of the ozone shield and global warming

# Study Activities

1. Explain the importance of the hydrogen bonding in water and how it is related to the biologically important characteristics of water.
2. Sketch a cross section of a typical lake or pond and indicate the major regions or zones.
3. List the major types or classes of organisms typically found in lotic ecosystems and indicate where they are located in the water column.
4. Compare and contrast the features of streams and rivers.
5. List at least five ecologically important functions of wetlands.

# 16

# Marine Ecosystems

## Objectives

After studying this chapter, the reader should be able to:
- Describe the role of tides, waves, and currents in marine ecosystems.
- Explain the general patterns of ocean currents.
- Describe the major zones of the marine environment.
- Define and give examples of the major marine biomes.

## I. Physical and Chemical Characteristics

### A. General information
1. Approximately 71% of the earth's surface is ocean
2. Salinity, currents, tides, and waves are major factors affecting ocean ecosystems
3. The mean depth of the oceans is approximately 3,700 meters; the deepest point is the Marianas trench in the Pacific Ocean, which reaches a depth of 10,750 meters

### B. Tides and waves
1. *Waves* are disturbances of the surface of water in the form of a moving ridge
2. Within waves, water oscillates back and forth without significantly changing its horizontal position, except along the shoreline, where waves break on the shore
3. *Tides* involve an alternate rise and fall of an ocean's water level generated by the effects of the gravitational pull of the moon and sun
   a. The rise (high tide) and fall (low tide) of tides occur across the oceans as the earth rotates and the waters are pulled alternately toward the moon and the sun
   b. The moon has a greater affect on tides than the sun because it is closer to the earth and its ability to cause tides is 2.25 times greater than that of the sun
   c. *Spring tide,* the largest tide of the month, occurs when the earth, moon, and sun are aligned (full and new moons) and the gravitational effects are additive and therefore strongest
   d. *Neap tide,* either of the two lowest high tides of the month, occurs when the moon and sun form a right angle with the earth; this orientation occurs at the quarters of the moon and causes a conflict between gravitational fields, minimizing the tidal effects

e. On a daily basis, the general period between high tides is 12 hours and 25 minutes, but this varies with local conditions

## C. Currents
1. *Currents* are produced by the interaction of surface winds and temperature differences in various parts of the ocean
2. High input of solar energy at the equatorial regions produces warm, expanding waters that tend to flow toward the poles
3. Prevailing winds and the earth's rotation distort the flow of ocean currents, forming a series of circular flow patterns or cells in the Northern and Southern Hemispheres
4. North of the equator, ocean currents generally circulate in a clockwise direction, with westward-moving currents just north of the equator and eastward-moving currents at about 40° N latitude
5. South of the equator, ocean currents generally circulate in a counterclockwise direction, with westward-moving currents just south of the equator and eastward-moving currents from 40° S latitude to the Antarctic

## D. Salinity
1. Higher evaporation in the tropical ocean regions results in greater salinity in the waters near the equator as compared to waters in temperate areas
    a. As saltier water flows poleward and cools, its higher density causes it to sink and results in a mixing of upper and lower depths of ocean waters
    b. Seasonal mixing of water from upper and lower depths may occur in shallow coastal regions of oceans
    c. In the open sea, mixing is restricted to the upper layers
    d. Much of the ocean has a permanent thermocline that varies with latitude and currents; it is closer to the surface in temperate areas and deeper in tropical waters
2. As a result of the reduced circulation between deep and superficial waters, much of the euphotic zone in the open sea is low in available nutrients
    a. Nutrients captured in the phytoplankton are carried below the thermocline and out of the euphotic zone when plankton die and their bodies drift downward through the water column and decompose at lower depths
    b. Distribution of marine zones is significantly influenced by patterns of nutrient flow into the euphotic zone

# II. Marine Zones

## A. General information
1. Marine ecologists recognize six basic marine zones: littoral, neritic, pelagic, benthic, upwelling, and coral reef (see *General Positions of Marine Zones*)
    a. The **littoral zone,** located where the ocean and land mass intersect along shorelines, generally is the area between high tide and low tide
    b. The **neritic zone,** found in that portion of the ocean covering the continental shelves, extends to a water depth of 200 meters
    c. The **pelagic zone** encompasses the vast expanse of open ocean; it comprises 90% of the ocean surface

## General Positions of Marine Zones

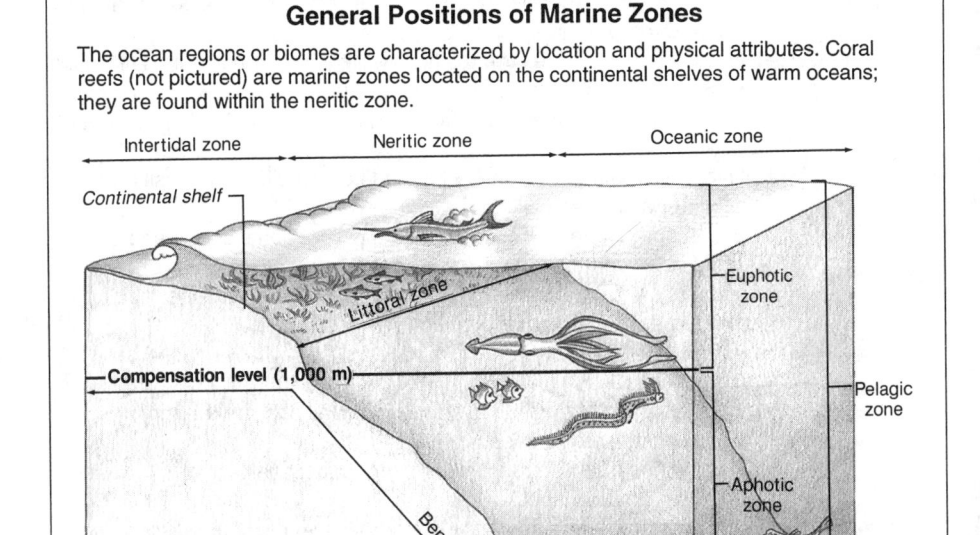

The ocean regions or biomes are characterized by location and physical attributes. Coral reefs (not pictured) are marine zones located on the continental shelves of warm oceans; they are found within the neritic zone.

- Intertidal zone
- Neritic zone
- Oceanic zone
- Continental shelf
- Littoral zone
- Compensation level (1,000 m)
- Euphotic zone
- Pelagic zone
- Aphotic zone
- Benthic zone

    d. The **benthic zone** comprises the ocean bottom below the neritic and pelagic zones
    e. The **upwelling zone** occurs where water currents are forced upward from depths to the euphotic zone
    f. **Coral reef zones** are widely distributed in shallow, warm seas of subtropical and tropical areas; as their name implies, they contain large coral structures built by carbonate-secreting organisms that inhabit the area
  2. Marine zones are significantly affected by patterns of nutrient availability in the euphotic zone
    a. Littoral and neritic zones receive nutrient input from adjoining continental areas and generally are rich in organic detritus and inorganic nutrients
    b. Upper pelagic regions tend to be low in available nutrients, whereas lower pelagic regions are enriched by detritus falling from surface waters
    c. The most productive marine ecosystems are concentrated along the continental margins and the upwelling areas, where nutrients from depths are flushed into the euphotic zone

## B. Littoral zones
  1. Located near the margins of continents, littoral zones are characterized by fluctuations in temperature, light, tides, and wave action

2. Littoral zones can be further classified as estuaries, tidal marshes, sandy shores and mudflats, and rocky shores
   a. **Estuaries** are located where streams and rivers join or empty into salt water
      (1) The waters of all streams and rivers eventually empty into the sea
      (2) Currents within estuaries are influenced by the currents of streams and tides
      (3) Salinity varies within an estuary because of the different densities of salt water and fresh water
      (4) Plankton and detrital-based food chains predominate in estuaries
      (5) Nutrient input from an entering stream or river may build up over the winter followed by a massive release in the spring, causing rapid growth of algae and other plankton
      (6) Estuaries often are very productive ecosystems with great species diversity
         (a) Salt marsh grasses, algae, and phytoplankton are the major producer organisms in estuaries
         (b) Estuaries also support a variety of worms, shellfish, crustaceans, and many fish species
         (c) Estuaries are vital breeding grounds for many marine fishes and invertebrates and also are crucial feeding areas for many semiaquatic vertebrates, especially waterfowl
      (7) Unfortunately, many estuaries (for example, Long Island Sound and Chesapeake Bay) are adjacent to major urban centers and thus receive large doses of pollutants and heavy use from the nearby human populations
   b. *Tidal marsh* ecosystems commonly form on alluvial plains around estuaries, sheltered bays, and islands; generally, they contain grasslike communities
      (1) At the water's edge, the mud or sand is colonized by algae, which later is replaced by eelgrass
      (2) As more sediment accumulates and the soil level rises above the water level, sea poa and cordgrass begin to grow
      (3) As the ground becomes firmer, marsh grass becomes established
      (4) Farther back from the water's edge, beyond the reach of high tide, various shrubs begin to grow
      (5) Salt marshes are among the earth's most productive ecosystems, as a result of the constant input of new nutrients caused by tidal action and deposition from streams and rivers
   c. Sandy shores and mudflats are littoral ecosystems found at the shoreline
      (1) *Sandy shores* and *mudflats* generally are barren habitat, with most of their life forms existing beneath the sand or mud rather than on it
      (2) These shore ecosystems exhibit zonation of vegetation and animal life related to tidal influences
      (3) Accumulation of organic matter in the substrate is necessary for community development; sources of this organic matter include detritus from seaweeds, decaying animals, and feces
      (4) The food chains of sandy shores and mudflats largely are decomposer-based
      (5) The individual sand grains commonly are coated with a gelatinous mixture of bacteria, algae, and particulate organic matter

(6) Animals typically found in these habitats include various amphipods, crustaceans, clams, nematodes and polychaeta, worms, mollusks, and sand dollars
(7) The primary difference between these ecosystems is the substrate; sandy shores are composed of granular sand particles ranging from 0.02 to 2 mm in diameter while mudflats are composed of silt and clay particles less than 0.02 mm in diameter
(8) Because of their dependence on imported organic matter and heterotrophic nature, sandy shores and mudflats are best considered as part of the entire coastal ecosystem rather than as separate ecosystems

d. The *rocky shores* of the northern Pacific and Atlantic coasts are characterized by forceful waves and tides
  (1) Many of the animals inhabiting the rocky coastal regions possess holdfast structures to maintain their position against the force of waves and tides; these sessile (attached) species are more prevalent on rocky shores than in any other biome
  (2) The rocky shore can be divided into three zones: an *upper spray zone* above the high-tide line but subject to saltwater spray, an *intertidal zone* subject to alternating submerged and dry periods by the changing tides, and a *subtidal zone* that is always submerged but subject to changing water depth and light intensity
    (a) The spray zone is characterized by an upper band of black, slimy blue-green algae often grazed by mollusks and a lower region inhabited by barnacles
    (b) The rocks of the intertidal zone generally are covered with brown and red algae, particularly *Fucus;* these algae are grazed by limpets
    (c) At the lower margin of the intertidal zone often is a dense growth of brown algae and numerous sea stars, crustaceans, and other invertebrates
    (d) The subtidal zone supports a variety of brown and red algae, as well as plankton, mollusks, barnacles, sea urchins, and other invertebrates

## C. Neritic zone
1. The neritic zone occupies the seas above the continental shelves and generally extends to a depth of 200 meters
2. Marine life is concentrated in the neritic zone because of the favorable nutrient levels
3. The major food chain in the neritic zone includes fish, copepods, sand eels, meroplankton forms of barnacles, mollusks, tunicates, arrowworms, and diatoms
4. Determining food chains in the neritic zone is difficult, because the consumers feed on various trophic levels and consume everything from herbivores to second-order carnivores
5. Most of the great commercial fisheries of the world are located on the continental shelves and upwelling areas (see "Upwelling Zones" below)
6. Waters of the continental shelves periodically experience marine algae blooms, which appear to be related to sunlight, water temperature, nutrient levels (par-

ticularly nitrogen and phosphorus), and biologic interaction of grazing zooplankton

## D. Pelagic zone
1. The pelagic zone encompasses the vast open seas and includes most of the ocean's water
2. Most of the pelagic zone is nonproductive, because food density and nutrient concentrations tend to be low
   a. Although the euphotic zone extends to a greater depth in the pelagic biome than it does in either the neritic or littoral biome, total primary productivity is much lower because of low nutrient availability
   b. In arctic and antarctic regions, the pelagic euphotic zones are higher in nutrients; consequently, species abundance increases
3. Plankton species are confined to upper layers of the biome, usually the upper 100 meters; zooplankton present in this biome include protozoa, copepod, krill, and jellyfish larvae
4. Fish and plankton of the region tend to be blue, making them difficult for predators to see
5. Whales are important consumers of the pelagic zone, feeding either on zooplankton or nekton organisms such as fish or krill
6. Other important nekton organisms include squid, fish, sea turtles, sea otters, sea lions, and porpoises
7. Pelagic sea birds, such as petrel, tern, albatross, and booby, catch fish in surface waters
   a. Diatoms and dinoflagellates are important primary producers in arctic regions
   b. Krill are the major herbivores in these ecosystems and are voraciously consumed by whales

## E. Benthic zone
1. The benthic zone of oceans is similar to the profundal zone of freshwater lakes, in that many nutrients reach the ocean floor by drifting slowly down from the upper regions of the water column in the form of detritus
2. The benthic zone near the coastal areas receives substantial sunlight; the amount of light and, subsequently, the temperature decline rapidly with increasing depth
3. The sea floor itself comprises sand or fine sediments (ooze) made up of silt and often the shells of dead microscopic organisms, such as diatoms
4. Neritic benthic communities are very productive
   a. These communities consist of bacteria, fungi, seaweeds and filamentous algae, sponges, worms, mollusks, crustaceans, echinoderms, and fishes
   b. Species composition varies with distance from shoreline, water depth, and composition of the sea floor
   c. Many organisms live buried in the sediment and soft substrate
5. Deep benthic communities consist of invertebrates and fishes and are found at great depths, where temperatures remain continuously low (approximately 3° C), water pressure is high, little or no light infiltrates, and nutrient availability is low

### F. Upwelling zones
1. Upwelling zones develop in areas where nutrient-rich deep ocean waters are forced to the upper euphotic zone, where the newly supplied nutrients fuel unusually high productivity
2. Large fish-eating bird populations may be associated with upwelling areas
3. In contrast to the marine abundance and productivity, the adjacent land area is likely to be coastal desert because prevailing winds blow from land to sea
4. The most famous upwelling biome is the Peruvian Anchovetas, located off the coast of Peru; here, southeasterly winds blowing along the coast draw surface ocean waters away from the coast, and the surface waters are replaced by an upwelling of deeper waters
5. The high productivity in upwelling biomes is subject to periodic disruption when winds shift
   a. Without the wind, upwelling slows or ceases, and the nutrient enrichment normally brought to the surface ceases
   b. Without the nutrient supply, fish die and fisheries collapse
   c. The shift in winds (and subsequent disruption of upwelling patterns) at the Peruvian Anchovetas is known as "El Niño" and is credited with causing fisheries to collapse

### G. Coral reef zones
1. Coral reefs are located in the warm tropical waters of the neritic zone
2. Currents and waves constantly renew nutrients to surrounding waters, and sunlight penetrates to the sea floor
3. Coral reefs are dominated by the corals, a diverse group of cnidarians that secrete a hard exoskeleton composed of calcium carbonate
4. The coral feed on microscopic algae and organic particles; they depend on the photosynthesis of symbiotic dinoflagellates that live within the bodies of the coral
   a. Organic compounds are transferred between the dinoflagellate algae and the coral, so both benefit from the association
   b. In addition to the symbiotic dinoflagellates, red and green algae commonly are embedded within the coral's calcium carbonate exoskeleton
5. A great abundance and variety of microorganisms, invertebrates, and fish live among the coral reefs

## Study Activities
1. Describe the mechanism by which major ocean currents are produced.
2. Sketch a diagram of the ocean and indicate the major zones or regions.
3. Describe the structure, location, and importance of an estuary.
4. Explain the concept of a marine upwelling and why it typically is a highly productive ecosystem.
5. Choose a major ocean zone and describe its major features and the types of organisms likely to be present.

# Appendix

# Selected References

# Index

## Appendix: Glossary

**Abiotic** — characteristic of the part of an organism's environment made up of nonliving things, such as soil, air, light, and nutrients

**Acclimatization** — process by which a range of adjustments occur in the body of an organism after exposure to natural environments in which a number of parameters vary simultaneously; parameters to which an organism may become acclimatized include temperature, photoperiod, humidity, and so forth

**Adaptation** — any genetically determined characteristic that improves chances of an organism transmitting genes to the next generation, increases the production of offspring

**Age structure** — distribution of individuals in a population into age classes

**Allogenic** — ecological change or development that is brought about by some factor(s) external to the ecosystem, such as fire or storms

**Alluvial** — term applied to regions of floodplains where sediments are deposited as water currents slow and particles settle out

**Altricial** — condition in which offspring are born or hatched usually blind and too weak to support their own weight; usually found among birds and mammals

**Amensalism** — relationship between two species in which one is inhibited or harmed by the presence of the other

**Ammonification** — breakdown of proteins and amino acids to ammonia by fungi and bacteria

**Aphotic zone** — that region of a water body below the depth of light penetration; usually from the compensation point to the bottom of the water body

**Autogenic** — ecological change or development that largely is controlled by factors within the ecosystem, such as when the vegetation alters microsite conditions by providing shade or adding organic matter to soil

**Autotroph** — organism able to synthesize required nutrients from inorganic substances found in its environment; autotrophs can use energy from the sun or from the oxidation of inorganic substances to make organic molecules from inorganic ones

**Benthic zone** — bottom region of a lake or sea

**Benthos** — organisms living on the bottom of a lake or sea

**Biochemical** or **biological oxygen demand (BOD)** — measure of the amount of oxygen needed to decompose the organic materials within a specified volume of water; the more organic matter, the more oxygen needed and the greater the BOD

**Biodiversity** — total number of species and genetic variety within each species

**Biogeochemical cycle** — movement of chemicals, compounds, or minerals through living organisms and the nonliving environment, generally considered at the biosphere level

**Biological magnification** — increase in concentration of slowly degradable, fat-soluble chemicals in organisms at successively higher trophic levels within an ecosystem; commonly occurs with metals, DDT, and PCBs

**Biome** — major, extensive and regional collection of plants and animals, often extending over vast geographical areas

**Biotic** — characteristic of the part of an organism's environment made up of other living things

**Biotic potential** — natural tendency or ability of a population to increase in size, denoted by $r$, and the difference between birth rates and death rates, also known as intrinsic rate of increase

**Calcification** — soil-forming process of dry climates in which salts are drawn to the soil surface as a result of excess evaporation of soil water

**Cannibalism** — special type of predator-prey interaction in which the predator and prey are of the same species

**Carrying capacity** — maximum number of individuals that can be supported in a given environment
**Circadian** — of or pertaining to the period of day and night within a 24-hour period
**Climax stage** or **community** — endpoint of ecological development or succession, characterized by a more or less stable community that is largely persistent and self-perpetuating
**Coevolution** — mutual influence of two species interacting with one another and reciprocally influencing each other's adaptation and evolution
**Cohort** — group of organisms of the same age living together
**Commensalism** — relationship between species in which one species benefits, but the relationship is either neutral or of no benefit to the other
**Community** — collection of interacting populations in an area; a naturally occurring collection of different species
**Compensation point** — that point in time and space where photosynthesis and respiration proceed at the same rate and no net gain or loss in carbohydrate occurs; in aquatic ecosystems, it is the line between the photic and aphotic zones
**Competition** — interaction between two species simultaneously utilizing the same resource
**Competitive exclusion principle** — hypothesis stating that when two or more species coexist and simultaneously use the same resource, one will displace or exclude the other
**Consumer** — any organism that lives on another organism, dead or alive; see also *Heterotroph*
**Coral reefs** — ocean zones of subtropical and tropical seas characterized by carbonate-secreting organisms; warm, shallow waters of the neritic zone
**Death rate** — number of individuals in a population dying during a given time interval divided by the number of individuals alive at the midpoint of the time interval
**Denitrification** — conversion of nitrates and nitrites to nitrogen by microorganisms

**Density** — in chemistry, the mass per unit volume of a substance; see also *Ecological density*
**Detritivore** — organism (other than bacteria and fungi) that feeds on dead organic matter
**Detritus** — partially decomposed plant and animal matter
**Dominant individual** — individual of a social hierarchy found at the top of the social order, with access to much of the resources available to the population
**Dominant species** — species that biologically controls a community by altering the availability or utility of resources
**Dystrophic** — term applied to a water body with high content of organic matter, often with high littoral productivity and low plankton productivity
**Ecological density** — number of individuals per unit of habitat or available living space
**Ecological release** — expansion of habitat or increase in food availability that results when a species is released from interspecific competition
**Ecology** — study of organisms in relation to their total environment; examines the interactions between organisms and between organisms and their environment
**Ecosystem** — system consisting of a biotic community and its abiotic environment
**Ecosystem stability** — measure of the integrity and resilience of an ecosystem that is subject to disturbance
**Ecotone** — area of intersection and transition between two or more ecosystems in which the ecosystems intergrade into one another
**Ectotherm** — organism whose body temperature is set primarily by external thermal conditions; includes all animals apart from mammals and birds
**Endangered species** — species in immediate danger of extinction because of extremely low populations
**Endotherm** — organism that regulates body temperature by internal (metabolic) heat production and loss

**Environmental resistance** — term used to describe the factors within an organism's environment that tend to decrease population growth rate by increasing death rate or decreasing birth rate

**Epilimnion** — upper, warm and oxygen-rich layer of a water body

**Estivation** — condition of dormancy in animals during a period of drought or a dry season

**Estuary** — protected and partially enclosed bay area where fresh and salt water meet and mix

**Euphotic zone** — that portion of an aquatic ecosystem from the surface to a depth where photosynthesis and respiration are equal

**Eutrophic** — term applied to a water body high in nutrients and productivity

**Evenness** — component of species diversity that addresses the equitability of distribution of individuals among the various species in a given area

**Fecundity** — number of offspring produced by an organism during its lifetime

**Fitness** — genetic contribution by an individual's offspring to future generations

**Food chain** — linear sequence of organisms, each feeding upon the lower trophic level

**Food web** — representation of energy flow through populations in a community; consists of many interconnected food chains

**Foraging efficiency** — measure of a predator's ability to concentrate activity in the most profitable areas of its habitat

**Functional response** — change in the rate of prey capture by a predator in response to a change in prey density

**Fundamental niche** — total range of environmental conditions under which a species can survive, considered in the absence of any biotic interactions

**Gene pool** — all the genes of all the individuals of a given population

**Gleyization** — soil-forming process that occurs in poorly drained areas, resulting in low oxygen concentration, slow decomposition of organic material, and development of characteristic gray color

**Handling time** — length of time needed for a predator to prepare or process and digest a prey item

**Heterotherm** — organism that becomes endothermic or ectothermic during various parts of its lifetime; for example, hibernating animals are ectothermic when dormant but become endothermic when active

**Heterotroph** — organism that depends for its survival on organic nutrients from its environment because it is unable to synthesize organic material; includes animals, fungi, many bacteria, and a few carnivorous plants; a consumer

**Hibernation** — state of dormancy entered into by many animals (including some arctic and temperate mammals, reptiles, and some amphibians) during winter in cold climates; metabolism is slowed and body temperature drops

**Home range** — area in which one is most likely to find an organism; area in which the species is located throughout the year

**Homoiotherm** or **homeotherm** — animal that maintains its body temperature by physiological means within a narrow range and often above the temperature of its surroundings, despite a wide variation in environmental temperature

**Horizontal heterogeneity** — spatial organization of a community such that species occur in a patchwork or mosaic pattern of clusters over the landscape

**Humus** — organic material derived from partial decomposition and decay of plant and animal matter

**Hypolimnion** — lower, cold, and oxygen-poor region of a water body that lies below the thermocline

**Invertization** — soil-forming process in which soils with a high clay content crack when dry; upper soil layers fall into these cracks during rain events and are transported downward through the profile

**Island biogeography** — study of the distribution and abundance of species on islands or other isolated habitats

**Iteroparous** — having multiple reproductive periods or efforts during a lifetime
**Laterization** — soil-forming process in which oxidation results in the formation of deeply weathered soils; occurs in hot humid climates
**Leaching** — dissolving and removing of nutrients by water out of soil, litter, and organic matter
**Lek** — small territory used by males of some species in courtship and mating
**Life expectancy** — average number of years to be lived in the future by members of a population
**Limnetic zone** — upper open water region of a lake or sea in which light penetrates to lower border of the water column
**Littoral zone** — region of a lake or ocean near the shore where light penetrates to the bottom, often with submergent and emergent vegetation present
**Marl** — deposit of calcium carbonate, clay, and organic material that can form in freshwater lakes with high calcium input
**Mesotrophic** — term used to describe an intermediate stage of aquatic succession marked by moderate amounts of nutrients
**Metalimnion** — transitional zone of a water body found between the epilimnion and the hypolimnion; region of rapid temperature decline with depth
**Mimic** — nonharmful species that imitates a harmful (model) species
**Mimicry** — relationship between species in which one (the mimic) benefits from a superficial resemblance to the other (the model)
**Model** — species in mimicry that often is unpalatable or toxic and is copied by the mimic; also, a representation of a natural phenomenon
**Monogamy** — state in which a single male pairs with a single female in a partnership that may last for an extended period, perhaps a lifetime
**Mortality** — death of individuals in a population
**Mortality rate** — number of individuals in a population dying during a given time period divided by the number of individuals alive at the beginning of the time period
**Mutualism** — symbiotic relationship in which two species benefit by enhanced survival, growth, or reproduction
**Natality** — production of new individuals in a population
**Nekton** — free-swimming aquatic organisms
**Neritic** — term applied to that portion of the ocean over the continental shelves, reaching a depth of 200 meters
**Neuston** — freshwater aquatic organisms that live at, near, or on the water surface
**Niche** — functional role of a species in a community; includes all of its activities, relationships, and requirements for survival and persistence
**Niche shift** — change in the behavior of an organism that occurs in order to decrease interspecific competition
**Niche width** — range of a single niche factor occupied by a population; also known as niche breadth
**Nitrification** — breakdown of organic compounds that contain nitrogen to nitrates and nitrites
**Nitrogen fixation** — chemical process by which gaseous nitrogen ($N_2$) from the atmosphere is converted to biologically useful forms (ammonia or nitrate)
**Numerical response** — change in predator density in response to changes in prey density
**Oligotrophic** — term applied to a water body low in both nutrients and productivity
**Optimal foraging** — strategy employed by a predator to maximize its rate of net energy gain
**Organic matter** — dead and decaying plant and animals material and waste found in soil
**Parasite** — organism that lives in association with and at the expense of another organism from which it obtains organic nutrients
**Parasitoidism** — interaction in which one organism attacks another (host) by laying its eggs in or on the host; the eggs hatch and the larvae feed on the host until it dies

**Parental investment** — amount of energy, time, or resources spent in the production, nurturing, and care of offspring
**Pelagic zone** — open water region of a lake or sea, including both the euphotic and profundal zones as well as the epilimnion and hypolimnion
**Periphyton** — freshwater aquatic organisms found attached to submerged plants, rocks, or debris
**Pioneer stage** — first stage in a sequence of ecological succession characterized by the arrival of the first plant and animals species
**Plankton** — small plants and animals found in freshwater or marine ecosystems, generally floating or weakly swimming
**Podzolization** — soil-forming process in which materials are leached from the A horizon and accumulate in the lower horizons; often iron, aluminum, silica, and clays are leached to lower depths
**Poikilotherm** — see *Ectotherm*
**Polygamy** — mating system in which one animal mates with several members of the opposite sex
**Population** — group of organisms of the same species living in the same area
**Precocial** — condition in which newborn or newly hatched offspring are more or less self-sufficient
**Predation** — relationship between two species in which one kills and consumes the other as a food source
**Predator satiation** — escape mechanism for prey in which reproduction is synchronized so that large numbers of offspring are produced in a short time, thereby allowing a sufficient number to escape predation
**Primary production** — assimilation or accumulation of energy by green plants and other autotrophs
**Primary succession** — development of a community on a new site never before occupied by living organisms
**Proclimax** — successional sere that remains or persists for an extended period during which succession appears to be arrested
**Producers** — photosynthetic green plants and certain photosynthetic or chemosynthetic bacteria that convert light or chemical energy into organismal tissue; organisms capable of synthesizing organic material
**Production** — accumulation of energy or biomass by an individual, population, or community
**Productivity** — rate of energy storage per unit time; often measured as $kcal/m^2/year$
**Profundal zone** — deep portion of an aquatic ecosystem, below the limnetic zone; also called sublittoral zone
**Psammon** — aquatic burrowing animals living within bottom sediments of a water body
**Realized niche** — that portion of the fundamental niche occupied by a population in the face of biotic interactions such as competition; environmental conditions under which a species lives in nature
**Relative abundance** — proportion a particular species contributes to the total abundance of a community
**Reproductive effort** — total of current and future reproduction of an organism; includes an individual's contribution to future generations in terms of producing offspring
**Richness** — component of species diversity that considers the total number of species present in an area
**Riparian** — term applied to areas associated with or adjacent to stream or river banks
**Salinization** — soil-forming process similar to calcification but more extreme; results in excessive accumulation of salts in upper region of soil profile as a result of water evaporation from soil surface
**Search image** — mental, sensory, or behavior construct formed by predators to enable them to find prey more quickly and efficiently
**Secondary succession** — progressive, step-wise development of communities on

a site that previously supported living organisms (for example, new forests established on abandoned farmland)
**Seed bank** — accumulation of dormant seeds that remain in the soil for extended periods
**Selection** — differential reproduction of one individual in a population as compared to other individuals; an organism has a selective advantage if it produces more offspring that survive to reproduce than are produced by other individuals
**Semelparous** — having only one terminal reproductive period or effort in a lifetime
**Sere** — one of many stages in the sequence of ecological development of an ecosystem during succession
**Sexual selection** — selection of a mate by female animals in which a particular characteristic or behavior is favored
**Social dominance order** — organization of a group of individuals into a distinct ranking in which higher-ranking individuals win aggressive encounters with lower-ranking individuals and thereby gain access to more resources within the population
**Species abundance** — commonness or number of species and the number of individuals in each species found in a community
**Subordinates** — members of a social hierarchy that rank lower on the social ladder and, therefore, sometimes are denied access to resources
**Substrate** — base or substratum on which an organism lives; for example, soil surfaces, rocks, tree trunks, or sea floors
**Succession** — replacement of one community by another, a process often ending in a stable community known as a climax community
**Survivorship** — number of survivors divided by the number of individuals alive at the beginning of a time period of interest
**Symbiosis** — condition in which two or more species live together in direct contact; see also *Mutualism*
**Territory** — area defended by an animal, often containing resources or requisites necessary for survival or successful reproduction
**Thermocline** — layer in a water body (within the metalimnion) in which temperature declines rapidly with increasing depth
**Threatened species** — species whose continued existence is in question because, although it still may be abundant in some parts of its range, its numbers have dropped drastically in other parts
**Threshold of security** — prey population density below which the prey is relatively safe from predation and above which predation is likely to increase
**Torpor** — state of inactivity entered by homoiotherms during periods of environmental stress; generally conserves energy and displays a characteristic decline in body temperature and metabolism
**Trophic level** — classification of organisms in an ecosystem according to feeding relationships; position within a food chain
**Upwelling** — nutrient-rich area within the ocean where deep water currents are forced upward to the euphotic zone
**Weathering** — mechanical and chemical breakdown or decomposition of rock material that leads to the formation of soil

# Selected References

Begon, M., Harper, J.L., and Townsend, C.R. *Ecology: Individuals, Populations and Communities* (2nd ed.). Boston: Blackwell Scientific Publications, 1990.

Caduto, M.J. *Pond and Brook: A Guide to Nature in Freshwater Environments.* Hanover, N.H.: University Press of New England, 1990.

Campbell, N.A. *Biology* (3rd ed.). Redwood City, Calif.: Benjamin/Cummings Publishing Co., 1993.

Clements, F.E. *Plant Succession: An Analysis of the Development of Vegetation.* Carnegie Institute of Washington No. 242, 1916.

Colinvaux, P. *Ecology 2*. New York: John Wiley & Sons, 1993.

Cox, G.W. *Conservation Ecology: Biosphere and Biosurvival.* Dubuque, Iowa: Wm. C. Brown Publishers, 1993.

Garrison, T. *Oceanography: An Invitation to Marine Science.* Belmont, Calif.: Wadsworth Publishing Co., 1993.

Gause, G.F. *The Struggle for Existence.* Baltimore: Williams & Wilkins, 1934.

Gleason, H.A. "The Structure and Development of the Plant Association." *Bulletin of the Torrey Botanical Club* 44: 463-481, 1917.

Jeffries, M., and Mills, D. *Freshwater Ecology.* New York: Belhaven Press, 1990.

MacArthur, R.H., and Wilson, E.O. *The Theory of Island Biogeography.* Princeton, N.J.: Princeton University Press, 1967.

Pianka, E.R. "On $r$- and $K$-Selection." *American Naturalist* 104: 592-597, 1970.

Ricklefs, R.E. *Ecology* (3rd ed.). New York: W.H. Freeman and Co., 1990.

Riechert, S.E. "The Consequences of Being Territorial: Spiders, A Case Study." *American Naturalist* 117: 871-892, 1981.

Smith, R.L. *Ecology and Field Biology* (4th ed.). New York: Harper & Row Publishers, 1990.

Stiling, P.D. *Introductory Ecology.* Englewood Cliffs, N.J.: Prentice-Hall, 1992.

# Index

**A**
Acacia tree, mutualism and, 97i
Acclimatization, 9
Accommodation, 9
Acid precipitation, 50
Adaptation, 9
Age structure, 56-57
Air
  circulation of, 11
  soil and, 26-27
Albedo, 11
Algal bloom, 146
Allelopathy, 72
Altricial animals, 87
Amensalism, 64, 100
Ammonification, 42
Animals
  light and, 21
  moisture and, 15
  nutrients and, 22
  temperature and, 17-19
Annuals, 89-90
Ants, mutualism and, 97i
Aphotic zone, 144i, 145
Atmosphere, 2
Autotrophs, 7

**B**
Bacteria, 44-45
Benthic zone, 144i, 145, 153i, 156
Benthos, 146
Biennials, 90
Biochemical oxygen demand, 142
Biodiversity, 108-111
Biogeochemical cycles, 38-45
Biological clock, 21
Biological magnification, 45
Biomass, 31-32, 36i, 37
Biomes, 2, 134-139
Biosphere, 2
Biotic potential, 57-58
Birth rate, 53
Bogs, 149-150
Broken stick distribution, 107
Buffering, 142

**C**
Calcification, 28
Cannibalism, 74, 85
Canopy, 102, 116i
Carbon cycle, 39, 40i, 41
Carbon dioxide, 48-49, 142
Carnivores, 7
Carrying capacity, 59
Chaparral, 135
Circadian rhythms, 21

Clements, F.E., 119
Climate, 10-13
Climate hypothesis, 134
Climatic stability theory, 105
Climax pattern theory, 113
Climax stage, 113
Coevolution, 95, 98i
Cohorts, 54
Cold hardening, 17
Cold stress, 16
Colonization, 122, 123i
Commensalism, 63-64, 99-100
Community
  biodiversity of, 108-111
  biological structure of, 103-107
  definition of, 1
  physical structure of, 101-103
Compensation point, 141
Competition
  interspecific, 70-73
  intraspecific, 64-70
Conduction, 16, 140
Connell, J.H., 119
Connell-Slatyer models, 119-121
Conservation, 110-111
Consumers, 7
Convection, 16
Coral reef zone, 153, 157
Counter-current exchange, 19
Currents
  freshwater, 142-143
  ocean, 12, 152

**D**
DDT, 46, 47i
Death rate, 54
de Candolle, A., 134
Decomposers, 7
Decomposition, 8
Defoliation, 78
Demography, 57
Density
  of population, 52-53, 65
  of water, 140
Dentrification, 42
Dentritivores, 35
Desert, 137-138
Detritus, 46
Dimorphism, sexual, 94
Dispersion, 53
Dominance, 66, 103, 104t, 107
Drought, 14, 127
Dystrophic lakes, 147

**E**
Ecological dominance distribution, 107
Ecological time theory, 105
Ecology
  experimentation in, 2-5
  levels of, 1-2
  models in, 5
Ecosystem
  changes in, 121-122
  components of, 6, 7i, 8
  definition of, 2
  disturbances to, 126-132
  freshwater, 140-150
  marine, 151-157
  stability of, 132-133
  terrestrial, 134-139
Ecotones, 102-103
Ectotherms, 17
Ehrenfeld, D., 110
Endangered species, 108
Endotherms, 18
Energy flow, 31-37
Environment, 1
Environmental resistance, 59
Epilimnion, 145
Equilibrium hypothesis, 123
Estivation, 19
Estuary, 154
Euphotic zone, 144i, 145
Eutrophic lakes, 147
Evaporation, 16
Evolutionary time hypothesis, 105
Exchange compartment, 38
Extinction, 61, 122, 123i

**F**
Facilitation model, 120
Factor compensation, 10
Fecundity, 89
Field capacity, 26, 44
Field studies, 3
Fire, 127, 129-132
Flooding, 127
Floodplains, 147
Flowers, coevolution and, 98i
Food chains and webs, 33, 34i, 35, 36i, 37, 47i
Foraging, 77-78
Forest, 102, 131, 138
Fossil fuels, 41
Freshwater ecosystems, 140-150
Functional responses, 75-77

**G**
Gaseous cycles, 39-44
Gene pool, 108-109

i refers to an illustration; t, to a table

# Index

Gestation, 57
Gleason, H., 119
Gleyization, 29
Global warming, 48-49
Grassland, 131, 135

## H
Handling time, 76
Heat exchange, 15-16
Heathlands, 136-137
Heat stress, 16
Heavy metals, 45-47
Herbaceous layer, 102
Herbivores, 7, 79
Hermaphroditism, 92-93
Heterogeneity, horizontal, 102
Heterotherms, 18
Heterotrophs, 7
Hibernation, 19
Homeotherms, 18
Home range, 69-70
Homoiotherms, 18, 33
Humans
 as danger to ecosystem, 127-128
 life table of, 55t
 ozone shield and, 48
Humidity, 13
Humus, 29
Hutchinson, G.E., 48
Hydrologic cycle, 13, 42, 43i, 44
Hydrosphere, 2
Hypolimnion, 145
Hypothesis, 3

## I-J-K
Immigration, 122, 123i
Inhibition model, 121
Insect infestation, 127
Invertization, 29
Island biogeography, 122, 123i, 124-125
Iteroparity, 58, 89
$K$ selection, 90, 91t

## L
Laboratory studies, 3
Lakes, 143, 144i, 145-147
Laterization, 28
Leaching, 23
Leaf area index, 8
Lek, 69
Lentic systems, 143-147
Liebig's law of the minimum, 10
Life expectancy, 54
Life table, 54, 55t
Light, 20-21, 48, 141
Limiting factors, 10
Limnetic zone, 144i

Lithosphere, 2
Littoral zone, 144i, 152, 153i, 154-155
Loam, 25
Lodgepole pine, 131
Log-normal distribution, 106-107
Lotic systems, 143, 147, 149
Lotka-Volterra competition equations, 70-72

## M
MacArthur, R., 90, 107, 122
Macronutrients, 21, 25-26
Marine ecosystems, 151-157
Marl lakes, 147
Marshes, 149-150, 154
Mating systems, 92-94
Mesotrophic lakes, 146
Metalimnion, 145
Microclimates, 12
Micronutrients, 21, 26
Mimicry, 81, 83
Moisture, 13-15, 26
Monoclimax theory, 113
Monogamy, 92
Mortality rate, 54-55, 56i
Mountains, 12
Mudflats, 154
Mutualism, 63, 96, 97i, 98-99

## N
Natality, 53-54
Nekton, 146
Neritic zone, 152, 153i, 155-156
Neuston, 146
Niche, 72-73
Nitrogen cycle, 41i, 42
Nitrogen fixation, 41-42
Nonequilibrium hypothesis, 133
Numerical responses, 77
Nutrients, 21-22, 146-147
Nutrification, 42

## O
Ocean currents, 12
Oligotrophic lakes, 146
Omnivores, 7
Organic matter, 26
Osmolarity, 15
Oxisolation, 28
Oxygen concentration, 141-142
Oxygen cycle, 39
Ozone shield, 47-48

## P-Q
Parasites, 8, 33, 64, 74
Parental investment, 87-88

Peatlands, 149-150
Pelagic zone, 144i, 145, 152, 153i, 156
Perennials, 90
Periphyton, 146
Permafrost, 139
Pesticides, 45-47
pH
 of soil, 26
 of water, 142
Phosphorus cycle, 45
Photoperiodism, 21
Photosynthesis, 8
Pianka, E., 90
Pioneer stage, 113
Plankton, 117, 146
Plants
 light and, 20-21
 moisture and, 14
 nutrients and, 21-22
 predation and, 78-80
 temperature and, 16-17
Pneumatophores, 14
Podsolization, 28
Poikilotherms, 17, 33
Pollination, 97, 98i
Polyclimax theory, 113
Polygamy, 92
Ponderosa pine, 131
Ponds, 143-147
Pools, 148
Population
 age of, 56-57
 competition and, 64-73
 definition of, 1
 density of, 52-53, 65
 extinction of, 61
 growth of, 57-59, 60i
 interactions of, 63-64
 mortality and, 54, 56i
 natality and, 53-54
 sex ratio of, 57
 survivorship and, 54-55, 56i
Precipitation, 43, 50
Precocial animals, 87-88
Predation, 64, 74-85
Preservation, 110-111
Proclimax, 114
Producers, 7
Production, 31-32
Productivity theory, 105-106
Profundal zone, 144i
Psammon, 146
Pyramids, ecological, 35, 36i, 37

## R
Rain forest, 138-139
Reasoning, 4-5
Reclamation, 128

i refers to an illustration; t, to a table

Reproduction
  effort in, 86-88
  patterns of, 89-90
Reservoir compartment, 38
Resilience, 122, 132
Resistance, 122, 132
Resources, conservation of, 110-111
Rete, 19
Rilles, 148
Rivers, 143, 147-149
r selection, 90, 91t
Runoff, 43, 127

**S**
Salinity
  freshwater, 140-141
  ocean, 152
Salinization, 28
Saprophytes, 7
Savanna, 130, 135-136
Scavengers, 33
Search image, 77
Seasonality, 21
Sedimentary cycles, 44-45
Seed bank, 54
Seiche, 143
Self-fertilization, 93
Semelparity, 58, 89
Sere, 113
Sex ratio, 57
Sexual selection, 93-94
Shade tolerance, 20
Shannon, C.E., 104
Shannon index, 104t, 105
Shelford's law of tolerance, 10
Shrubland, 136
Shrub layer, 102
Simpson's index, 103, 104t
Slatyer, R.O., 119
Social dominance, 66
Soil
  classification of, 30
  components of, 23-25
  formation of, 27-29
  profile of, 24i
  properties of, 25-27
Solar radiation, 11, 31
Spatial heterogeneity theory, 105
Species
  abundance of, 106-107
  biodiversity and, 108-111
  distribution of, 20
  diversity of, 104t, 105-106
  dominance of, 103, 104t
  endangered, 108
Spodsolization, 28
Stability-time theory, 106
Stomata, 14

Stratification
  of plants, 101-102
  of water, 141
Streams, 143, 147-149
Stress, population density and, 65
Sublittoral zone, 144
Substrate, 117
Succession
  allogenic, 112
  aquatic, 117, 118i
  autogenic, 112
  island biogeography and, 122-125
  models of, 118-121
  primary, 112, 114-115
  secondary, 112, 116i, 117
  stages of, 113-114
  terrestrial, 114-115, 116i, 117
Sulfur cycle, 44-45
Survivorship, 54-55, 56i
Swamps, 149-150
Symbiosis, 63-64, 96

**T**
Taiga, 138
Temperature
  animals and, 17-19
  plants and, 16-17
  species distribution and, 20
  survival and, 19
  thermal exchange and, 15-16
  water and, 140, 145
Terrestrial ecosystems, 134-139
Territoriality, 66-69
Theory, 3
Thermal exchange, 15-16
Thermocline, 145
Thermogenesis, 17
Threshold of security, 76
Tides, 127, 151-152
Tolerance model, 120-121
Tolerance range, 9
Torpor, 18-19
Toxins, 45-47
Trade winds, 12
Trophic levels, 34-37
Tundra, 139

**U**
Ultraviolet light, 48
Understory layer, 102, 116i
Upwelling zone, 153, 157

**V**
Validation, 5
Variables, 4

Vegetation, layers of, 101-102
Verification, 5

**W-X-Y-Z**
Water, 13, 42-44, 140-157
Water table, 44
Water vacuoles, 14
Waves, 127, 151-152
Weathering, 27
Weaver, W., 104
Wetlands, 149-150
Wildfire, 129
Wildlife
  fire and, 132
  preservation of, 110-111
Wilson, E.O., 90, 122
Wind, 12, 126

i refers to an illustration; t, to a table